Yes You Can! Help Your Kid Succeed in Math

Even if You Think You Can't

by

Jean Bullard   Louise Oborne

Yes You Can! Press

Seattle, Washington

© 1997 by Yes You Can! Press (A Division of Bufflehead Publishing, Inc.)

All rights reserved. No part of this publication may be reproduced or transmitted in any form by any means, electronic or mechanical, including photocopy, recording, or any information storage or retrieval system, or otherwise, without permission in writing from the publisher.

Care has been taken to trace ownership of copyright material contained in this text. The publisher will gladly receive any information that will rectify any reference or credit line in subsequent editions.

Published by Bufflehead Publishing, Inc.
2400 NW 80th Street, Suite 173
Seattle, Washington 98117

Cover Design: Steve McEachern, Quack Communications
Graphic Design and Desktop Publishing: Brenda Robson
Initial Input: Ray Claxton
Copy Editor: Cynthia Young
Illustration: Pat Cupples

Library of Congress Catalog Card Number: 97-60736
ISBN: 0-9658044-0-2

Printed and bound in the United States of America
by Patterson Press, Benton Harbor, Michigan

About this Book

Who's this book for?

It's for any adult who wants to help a child who is having difficulty with math. It doesn't matter how well—or how poorly—you did in math when *you* were a kid. With the assistance of this book, you *can* help. And you can do it without damaging your relationship with your kid. In fact, we'll show you how collaborating with your child can enhance your relationship.

Who wrote it?

Jean Bullard is a former math teacher, now author, editor, and education consultant. Louise Oborne is a successful author, editor, and communications consultant. She's also a former math dropout. This experience makes Louise uniquely qualified to co-write a book like the one you're holding.

Why?

This book grew out of our experience as we co-edited a series of math books for a major publisher. Jean had to talk Louise into working on the project. Louise thought she couldn't do it because she didn't do well in math in school. Jean thought that experience would be of enormous benefit to the project, and it was. Why? Because Louise asked questions about every line that didn't make complete sense— the same questions a student might wonder about, but not know enough to ask. As we worked together, we began to realize how this approach might work for parents who think they aren't good enough to help their own kids. That's why we decided to write this book.

Why is it so thin?

We don't believe that bigger always means better. In our minds, a big book may be a sign of poor editing. It forces the user to sift through page after page searching for a useful kernel of information. This book is "presifted." We've done the sorting for you so you can start helping your kid succeed in math as quickly as possible. A glance through the contents on pages 4 and 5, and the introduction on pages 6 and 7 will help you find the information you need to get started.

Done with empathy, coaching your child in math can be a satisfying, unifying experience that gives both of you increased confidence. We hope this book will help many thousands of parents achieve that goal.

Contents

Introduction
    Feeling Bad Because You Think You
Can't Help Your Kid in Math?.................................6
How this book can help

Section 1
    Pep Talk 9
Motivation, inspiration, and stimulation for adults who'd like to help their kids with math but think they can't

Chapter 1   Let Yourself Off the Hook........................... 10
Why bad math feelings happen to good people

Chapter 2   What Your Kid Is Up Against 12
Factors beyond your kid's control may be causing problems

Chapter 3   What Your Kid's Teacher Is Up Against
—and How It Affects Your Kid 14
Factors beyond the teacher's control may be causing problems

Chapter 4   You as Coach—What You Can Do,
How You Can Do It...................................... 16
You are good enough—your personal strengths can help you

Chapter 5   How to Get Your Kid to Buy In 18
Selling your kid on the value of collaboration

Section 2
    Winning Strategies..................................... 21
Tactics, techniques, and targets

Chapter 6   How to Start Coaching and When 22
Don't wait until next season—there's no time like now

Chapter 7   Handling Homework—Keeping Up the Daily
Grind without Getting Ground Down 24
The warm body hypothesis

Chapter 8   Focusing on Fundamentals 26
Getting over learner resistance and choosing the right foundation topics

Chapter 9   Teaching Your Kid to Walk and Talk Math 28
Creating a math-rich environment

Chapter 10   Plugging into Math—How to Get the Current
Flowing 30
Unblocking the learning circuit

Chapter 11   Making Sense of Math—The Power of Reading 32
Reading strategies to help your kid make sense of math

Chapter 12	Getting Your Kid to Talk—The Role of Verbalization in Learning Math .... 34 *Thinking out loud can pay big math dividends*	
Chapter 13	The "Write Stuff"—The Role of Writing in Learning Math .... 36 *Writing about math ideas can help your kid think more clearly*	
Chapter 14	How to Use a Textbook as a Coaching Manual .... 38 *What a textbook can—and can't—do for you*	
Chapter 15	How to Develop Helpful Habits and Routines .... 40 *The learning log habit and developing effective routines*	
Chapter 16	How to Recognize and Reward Success .... 42 *The value of positive feedback and short-term targets*	
Chapter 17	Tackling Test Fear—How to Coach Your Kid to a Personal Math Best .... 44 *Preparation, positive imaging, and trial runs lead to success*	
Chapter 18	Celebrate Success—How to Turn Tests into a Positive Experience .... 46 *How to make sure your kid gets maximum benefit from writing tests*	
Chapter 19	What if You Have to Turn Your Coaching Job Over to Somebody Else? .... 48 *Getting the best value without taking out a mortgage*	

Section 3

	Coaching Secrets .... 51 *Professional secrets for the at-home math coach*	
Chapter 20	Breaking the Secret Code.. 52 *A translation guide for the math coach*	
Chapter 21	The Secret Behind Addition and Multiplication ..... 54 *The secret revealed—multiplication is accelerated addition*	
Chapter 22	The Secret Behind Subtraction and Division .... 56 *Understanding how your kid can get done in by division*	
Chapter 23	The Secret Behind Calculators .... 58 *Outsmarting the robot in the calculator*	
Chapter 24	Secrets for Re-energizing Drill .... 60 *How to beat the times table blues and give your kid a boost*	

Selected Resources 62
Resources from Yes You Can! Press 63
Ordering Information 64

Introduction

Feeling Bad Because You Think You Can't Help Your Kid in Math?

Karin got on an elevator in a downtown hotel where a large business convention was taking place. The elevator was crowded with convention people who were heading to a cocktail party. They were cheerful and talkative and greeted Karin enthusiastically. An impeccably dressed man in his late thirties turned to Karin and smiled broadly.

"Are you here for the convention?" he asked.

"No," Karin replied. "I'm here for a teachers' workshop."

"Oh! What do you teach?" asked the man.

"Math," said Karin.

The elevator fell silent. There were eight floors to go.

Help Is on the Way!

Karin's story is true. It shows the power that one word—*math*—can have over people. It can make otherwise rational adults feel like social misfits. It can shake the confidence of the heartiest souls. But why? Because many people feel really bad about math. Some feel ashamed because they failed math in school. Others feel intimidated because they think they have forgotten everything they ever knew! Are you one of these people? Have you spent years perfecting math avoidance techniques? If so, you are in plentiful company. All around you are people who feel bad about math decades after their last math class. Like you, these people feel paralyzed at the prospect of helping their kids with math. They not only feel bad, they feel hopeless.

How this Book Can Help

We have written this book because we know there is hope. We believe wholeheartedly that you can help your kid in math even if you think you can't. You do not have to be a math teacher or a math genius. We will show you there is an alternative—you can be a math coach! Our experience—Jean's as a math teacher and Louise's as an editor of math textbooks—has convinced us that what you need most is confidence. And coaching can boost your confidence. We know that by using the strategies in this book, you will be able to help your child in ways you hadn't thought possible.

What You'll Find Inside

This book is for the adult helper, not for the child who needs the help. It will give you the tools and techniques you need to help your kid meet everyday math challenges head on. It is divided into three sections.

Section 1: Pep Talk

Regardless of your own math ability, this section is for you. Those who don't feel competent in math will find support and reassurance. You are good enough! Those who do feel competent may be tempted to skip this section. Why read it if you don't have any bad math feelings? Answer: *Because it can help you understand why your kid does.* You must understand this to persuade your child to collaborate with you. Empathy is essential if the barrier is to be breached.

Section 2: Winning Strategies

Besides providing general advice on coaching strategies, this section shows how exploring where your child's math circuits may be blocked can lead to strategies for getting the current flowing again. As well, you will learn how to help your kid prepare for "the big test," do the best possible job on it, and go on getting benefits after the test is over.

Section 3: Coaching Secrets

In math, the simplest sounding words can confound a struggling student or amateur coach. Terms such as "number" and "operation" may not mean what you think. And the traditional order for teaching certain topics may not be the best order for learning them. This section reveals what standard math textbooks never deal with.

Think of this book as a portable coaching clinic. Flip through it to get an idea of the range of material that is available. Sample the "sessions" that seem most applicable to your child's situation. In some cases, you'll find a cross-reference to another chapter that might provide additional guidance. As you familiarize yourself with the book, you'll find yourself coming back again and again, and getting more out of it every time you open it.

Section 1
Pep Talk

Chapter 1 Let Yourself Off the Hook

Alaina is an intelligent, energetic woman. She runs a successful landscape design business and is a single parent. Recently her 13-year-old daughter Emma, who normally has good grades, has begun to fail math. Emma gets defensive and angry whenever Alaina tries to talk to her about math. And Alaina feels frustrated and guilty because she failed math in high school, too. "How can I help Emma?" Alaina comments to a close friend, "I can't even keep my own books!"

Why Bad Math Feelings Happen to Good People

If you cringe at the very thought of math, this chapter is for you. We will help you understand where those feelings come from and why your lack of confidence isn't your fault. If the idea of math doesn't make you yearn for an out-of-body experience, this chapter will help you understand why other adults—and your kid—do.

As you read through our list of factors that may have led to Bad Math Feelings, mentally check off the ones that hit home for you. Then look for those that may apply to your child. Are they the same?

Too much to chew

When you were in school there was just too much math to teach. Teachers were expected to cover a vast amount of material each year and you were expected to digest it all. How many people at a banquet are able to eat everything they're offered? Not many. Inevitably many topics got crammed into the last few weeks of school. If you remember math class at the end of a school year— and many of us don't!—you probably weren't focused on what your teacher was saying. And even if you were paying attention, your plate was already full.

Practice, practice, practice

Because there was so much to teach and so little time to teach it, you probably didn't get enough time to practice math to absorb new ideas. And if you had a question about an exercise, your teacher probably didn't want to take up precious class time explaining it to you. Think back. How motivated were you to stay after school to do math? If you didn't have time to practice, you wouldn't have remembered new ideas. And remembering math ideas is critical to building a foundation for new learning. Practice may not make perfect, but it helps build self-esteem.

Shakin' all over

You may have been hurt by the cumulative nature of mathematics. In math, cumulative means that new knowledge builds on previous knowledge. To become confident in math, you needed to understand everything "from the ground up." If you missed class on Wednesday, then Thursday's lesson may have sailed over your head. This would have made your math foundation pretty shaky.

Out on a limb

Kids have always been unwitting subjects in educational experiments. You too may have been a victim of certain teaching styles that were in vogue. If you were in school in the 70s, you may have been made responsible for your own learning. Do you remember working alone or in small groups with minimal supervision and support? This was a great privilege for many kids. Real freedom! Like being told you could climb a tree. But you may not have recognized that you were getting into trouble. And if even if you did, you may have been too embarrassed to ask for help. The hardest part of tree climbing, of course, is getting down.

Left in the dust

Your family may have moved at a critical time in your math education. You may have lost time during the move, especially if it was across the country. Your new school may have been using a different textbook or even a different curriculum. And your new classmates may have thought you were weird because you didn't understand stuff they took for granted. You may have been left in the dust.

The failure trap

Did you ever flunk a math test? If so, you were bound to feel anxious about the next test. If you expected to do poorly, you probably did. This set up an expectation of failure that became a pattern. You internalized an image of yourself as a "math dummy." At this point, you may have fallen into the "math isn't cool" trap—a good strategy to save your self-esteem.

Math is for nerds

Were you great at phys ed, art, or English? Did you wonder why on earth you would need math to become an artist or an athlete? Then you probably thought math was for nerds. The relationship between math and your everyday life was not promoted when you were in school. It wouldn't have been evident to you then that having confidence in math would be a valuable asset as an adult.

You're More than OK

The math experiences you had in school happened to your "kid" self. Like all kids, you were subject to raging hormones, peer pressure, and distractions galore. You are no longer the kid whose face turned scarlet when asked to solve a problem at the board. But it's amazing how many adults have the same response today when confronted with a math "situation." It's time to let your "kid" self off the hook and let your "adult" self swing into action. You have far more skills today than you did when you were fumbling at the board in grade 9. You've had years of life experiences that have made you a caring, competent adult. These are the experiences you can draw on to help you help your kid. And you can help your kid. Just read on.

Chapter 2   What Your Kid Is Up Against

It's 9:05 a.m. and Manuel has just settled, breathless, into his seat. His teacher is already handing out the morning math drill. Manuel's stomach is growling. His baby sister is sick and Manuel has been taking care of her while his parents run the family store. Manuel didn't have time for breakfast this morning. He was trying to do the math homework he couldn't do the night before. Now Manuel looks down at the drill sheet. It looks like there are a thousand questions! The teacher holds up his hand, fingers spread wide. "Five minutes!" he shouts. Manuel swallows hard. His mind is a blank.

Morning Drill Torture

Manuel was able to answer only a few questions before he had to pass his drill sheet on to another, presumably "smarter," kid to mark. By 9:10 a.m. Manuel had already failed a test. You probably experienced similar situations when you were in school—almost certainly your kid has. The result is often anger and frustration—and all before lunch! But the good news is that your child may not have internalized these failures yet. There's still hope! You can still help salvage your kid's self-esteem and math potential if you know what she or he is up against.

Morning drill is just one of many challenges faced by your child every day at school. It's not the fault of the teacher. Often drills and quizzes are the only way a teacher of a large class can get an inkling of what kids know. Many teachers today feel that kids face much tougher obstacles than you did when you were in school. Remember how you felt back then? Well, it's much more difficult for kids today. Here's why.

Babies no More

You only have to spend a couple of hours watching television with your child to realize that issues you faced as an adolescent—sex, drugs, peer pressure, hormones—are faced by kids at a much younger age today. Peace, love, and tie dye have given way to AIDS, global warming, and joblessness. Make no mistake. Your kid knows a lot about these things. But like kids of any generation, kids today are struggling to be independent. And like you at their age, it's hard for them to make informed choices about their own long-term interests—especially if decisions have to do with their math future. It's the farthest thing from their minds. You can count on it.

Bigger not Better

If there was too much math to teach when you were a kid, there is even more now. The math curriculum your kid is up against is bigger than ever. It's also harder than ever. Even after years of consideration by committees all over the country, many topics on the math curriculum are taught two years too soon for the average kid. No wonder your child can't keep up! And math isn't the only subject your kid has to deal with. Other subjects have hefty requirements. If math homework takes too long or is too frustrating to do, your child may push it aside in an effort to keep up in other subjects and avoid the frustration of trying and failing.

Not All Teachers Are Created Equal

Many math teachers are specialists. They had no trouble in math when they were in school. They easily learned how to interpret symbols and mathematical shorthand such as equations. These teachers may have a hard time understanding anyone—like your kid—who doesn't have the same abilities. These teachers find it difficult to help kids who need to learn math through reading, spoken language, and pictures rather than through symbols and numbers. They may even think kids with math "problems" are lazy, stupid, or just daydreaming. What a specialist teacher may fail to see is that your child might just be hypnotized by incomprehension.

> **House of Cards Syndrome**
>
> The only way kids can get a grip on math is by having a solid foundation to support new knowledge. If your kid is having trouble, you can be certain that he or she is suffering from house of cards syndrome—one more missed concept and everything will collapse. In no other subject do kids need to know everything that went before to understand today's lesson. No wonder the foundation has to be strong! Your child probably knows this already, but has found it's impossible to shore up the foundation without working on the upper stories at the same time. This where you can help. We'll show you how in Section 2.

Given the trend toward budget cutting and staff downsizing in schools, some math classes are taught by teachers who have no math background at all. A teacher who does not have a thorough understanding of math concepts won't be able to communicate with kids who are having trouble.

We don't want to beat up on teachers. Their jobs are pretty tough. These two examples illustrate two extreme teaching situations. Most math teachers fall somewhere in-between. They know the math and they have a pretty good idea of how to teach it. But they are up against difficulties that prevent them from doing their jobs the way they would like to. In Chapter 3 you'll see why.

Chapter 3: What Your Kid's Teacher Is Up Against—and How It Affects Your Kid

John teaches middle school math. Each class is in a different room. His typical lesson plan includes 5 minutes on roll call and other business, 20 to 30 minutes teaching a new lesson, 10 minutes answering questions, and 15 minutes helping kids start their homework. Today three students John has already marked absent come in late. John stops the lesson and changes the attendance record. He settles the class down and starts the lesson again. But the teacher who last used the room took the last decent piece of chalk. As John hunts through Ms Lopez's desk, he can see Thomas and Bo pass something back and forth. John clears his throat—he feels suspicious of all his students these days.

John launches into the lesson again. Now Sabrina asks to go to the washroom. By the time she returns, John is part way through an example. Sabrina shoots up her hand to ask a question about it. John has only 15 minutes left and he has covered only part of the lesson. He races through the rest, hoping there will be time for kids to ask questions. There isn't. The bell rings. John barks out the homework assignment as the kids stampede out the door. John is exhausted. And he feels bad because he knows Kiel and Susan are having real problems. But there isn't any time today. There never is.

The Challenge of Teaching

There are remarkable teachers in America. Many are dedicated, enthusiastic educators whose value to our kids is incalculable. But even they have shared John's experience. In a perfect world, a 60-minute class with 30 kids might give each kid two minutes of personal teaching time. Not much, is it? And a teacher in John's situation couldn't spare even this. So if your kid is having trouble, the chance of one-on-one time with his or her teacher is slim at best. At worst, the teacher may not even be aware that your child is struggling.

Large classes are not the only obstacles faced by teachers. The whole teaching experience has changed since you were in school. Teachers have little input into what they are required to teach. They enjoy little respect—not only from their students but also from their colleagues and their boards. Because of cutbacks, teachers often find themselves teaching not only their own subjects, but those they may know little about. And these are just a few examples of what your kid's teacher may be up against. There are more.

Teachers as Parents, Teachers as Cops

Teachers today are up against a change in "clientele." Compared to 20 years ago, teachers face ever-increasing numbers of kids who suffer from poverty, hunger, and behavioral problems. Teachers of early grades find they need to raise children before they can teach them. Teachers of later grades are confronted with kids who have short attention spans, discipline problems, and even violent tendencies. It's hard to be effective when you're scared of your students. Teachers who spend the bulk of their time on "crowd control" have very little time or energy left to help kids who are floundering.

Curriculum Rich

Math teachers are up against the unrealistic expectations of curriculum gurus who set goals that may be impossible to achieve in a real classroom. A typical curriculum guide for all high school subjects used to be no more than 32 pages. Now the guide for one midsize American state is over 300 pages! What teacher has time to read such a document, let alone teach from it?

Resource Poor

Teachers are up against inadequate resources. For example, the latest curriculum trend is "interactive" math. Interactive math means kids interact with real situations using real materials and equipment such as sugar cubes and bathroom scales. But no money is available to buy the learning aids teachers need to make this work. Some teachers spend hours making their own "props" such as fake money. Some buy the stuff they need out of their own pockets. And some give up.

> **Filling the Gap**
>
> No matter what your math ability or your educational background, you can provide a better learning experience for your kid. Coaching doesn't require magical powers or expert knowledge. Your life with your child is rich in experiences in which you have already been a coach.
>
> Think about the last time you took your kid to a zoo or museum. You're probably not an expert on giraffes or dinosaurs. But your child probably turned to you for some kind of interpretation about what she or he was seeing. And without thinking, you probably were able to help your child make sense of an interpretive sign or map. That's coaching! And it can go a long way to fill the gaps that even a good teacher can't fill. In Chapter 4 we'll show you more about how you can be the best coach your child could have.

The Stop-Gap Solution—Destructive Teaching Techniques

Do you see how all these factors—crowded classrooms, exhausted teachers, unrealistic expectations, difficult kids—can lead to destructive teaching techniques? Morning drill is only one example. There are others. Mrs. Faro teaches a lesson one way, one time. When a child asks for help, she "reteaches" the lesson in the same way. But the kid doesn't get it the second time either. And Mrs. Faro doesn't have time to think up another approach. In the chaotic atmosphere of today's classroom, there is time only for the "right" answer. And it has to be fast. Kids get little credit for effort. And worst of all—they get no time to think.

Chapter 4

You as Coach—What You Can Do, How You Can Do It

Rosalie knew nothing about wrestling. But she did know that her son's wrestling team needed a coach. They lost more matches than they won. And they were discouraged. Rosalie volunteered. She went to her first practice armed with a whistle and a plate of chocolate chip cookies. "If you practice hard today," she told the boys, "you get the cookies. It's that simple."

Rosalie encouraged, coaxed, listened, and supported. She soothed sore limbs and raised flagging spirits. And she baked carloads of cookies. By the end of the season, Rosalie still didn't know much about wrestling. But her team got to the semifinals. When asked how she did it, Rosalie smiled. "It could have been the cookies," she joked, "but mostly it was the boys."

The Best Person for the Job Is You

Rosalie's cookies were a symbol of love and support—two powerful ingredients in the coaching "recipe." We think you have these ingredients in abundance. And that's what makes you the best math coach your kid could have. When you are supportive and positive, your child's self-esteem and math potential can still be saved. If you had trouble in math, you probably know better than anyone where your kid is having problems. Think about how you felt. Your child probably feels the same way. This puts you on the same team, with you as coach. The real challenge for you is to set realistic goals and know your role.

Understanding vs. Survival

All modern math curriculum guides emphasize the importance of applying math to everyday life. They would make you think your coaching goal should be understanding. But in Chapter 3 we said that curriculum guides often set up unrealistic expectations that neither teachers nor students can meet. Because so many math topics must be taught too soon for the average kid, understanding is often an unrealistic goal. Many kids have already suffered math "trauma." Their confidence is shaken and they are reluctant to invest any more of themselves in trying to understand math. These kids may think they will never understand math, so why try. Your child is probably one of them.

As your kid's coach, achieving understanding may not be your main objective. You may decide that what you really want is for your child to pass math and to be less miserable about it. These are survival goals. And survival is different from understanding. We're not saying that understanding is impossible. Understanding often comes after kids learn to survive certain math situations. Sometimes all it takes is getting a right answer. Eventually your child may see a pattern—that leads to understanding.

Coaching vs. Teaching

You don't need to become a math teacher. Your child already has one. Your role is math coach. Remember that a coach doesn't have to be an expert player. You're probably not a professional cyclist, right? But that didn't stop you from acting like a coach when you helped your kid learn to ride a bicycle. Some kids learn this skill easily. Others need more support—a hand on the back of the bicycle seat. Like any good coach, you provided the level of support your kid needed. You can do the same thing for math.

Advantage vs. Perfection

If your child is struggling in math, you can't expect to turn him or her into an A student overnight. Your first goal should be to guide your kid through math with minimal damage to his or her self-esteem. This alone would be a major achievement. Your child may never achieve an A in math. Your child may only get a C. But passing, if just barely, opens doors. Your kid may begin to appreciate that there are more "perks" attached to passing than to failing—your applause is one. This could spur your child toward improved math skills and end the losing streak. That's what coaches do best isn't it?

In Section 2 we'll show you some winning strategies to help you help your kid. But first read Chapter 5. It will show you how to get your kid to join you on the winning team!

Chapter 5   How to Get Your Kid to Buy In

Kim is a single parent. He wanted to find some way to show his son Brendan how important math is. No amount of talking made an impression on Brendan. He was convinced he didn't need math to be a professional basketball player. So Kim registered in a math course himself. In the evenings, after they had done the dishes, Kim and Brendan sat down at the kitchen table and worked on their math homework together. Brendan continued to moan about math, but after a month he realized that it was easier to talk to his father about some of the problems he was having simply because Kim was there. Kim didn't do Brendan's math for him—he had too much of his own! But Brendan's math grades began to improve and so did his relationship with his father.

Join the Same Team

Some kids may be desperate enough to cling to any support you offer. Others may not want you—or anyone—to know about their math "shame." These kids may have built a wall against despair. Demolishing the wall is not your job. But joining your kid on his or her side of that wall is. Because of parent-child conflict in other areas, your kid may balk at you "horning in." So how can you build a conflict-free partnership with your kid? By establishing that you and your child are on the same team. That's what Kim did. You don't have to take a math course to show your commitment. But you do have to demonstrate that you're not ganging up on your kid and you're not in cahoots with the math teacher. This struggle is about you and your child against math. In at least this one area you both have everything to gain by becoming partners. But it takes time.

Partnership Is Possible

Although it may not seem like it now, you have been a partner in your child's learning throughout his or her life. Remember when your kid learned to walk and talk? You praised every step, sympathized with every spill, and responded with enthusiasm to every word. You can do the same thing to help your child develop survival skills in math.

Active support, even when the going gets tough, shows that you have made a commitment and are willing to live up to it, even if you don't know all the answers.

Your Kid's Got What It Takes

Getting your kid to buy in may mean that you have to remind your child of the many times she or he struggled to do other tasks and succeeded—riding a bike, doing a somersault, building a model, learning to swim, batting a ball.

Try starting your "pitch" by saying something like: "Remember when you had a whole list of spelling words to learn and you could only get four or five right? We figured out tricks to remember the spellings of hard words. And you stuck with it and got 13 out of 15 on your test? Well, I think we can do the same thing with math!"

Keeping Your Kid on the Team

Helping kids feel good about themselves and part of a team is a critical coaching goal. Convincing your child that you are both on the same team is a tough assignment. It requires continuous communication and encouragement. Here are some strategies you can use to help your child build self-esteem and be aware that the two of you really are teammates.

Communication means taking the time to listen to your child and involving her or him in developing survival skills and strategies to conquer math fear.

- Demonstrate that you're not imposing a police state—you're just trying to support your child's struggle to survive in math.
- Involve your child in setting up a joint schedule for math activities. You should both realize that achieving the goal of survival will cost you time and attention. As a parent, you may have already gained some credibility by helping out at school, club, or church activities. So your kid knows that if you "sign up for the season" you will live up to your commitment.
- Involve your child in establishing realistic survival goals. For one kid, success might mean better marks than last year. For another, success might mean actually passing math.

Encouragement means "building courage"—focusing on your child's strengths and efforts so that he or she wants to keep on trying.

- Keep track of improvement. Write down math successes.
- Acknowledge discouragement. Never punish. Discuss problems and ask: "How do we get over this so we can win?"
- Encourage every step. "Last week you knew six of the 7 times table. This week you know nine."
- Accentuate the positive. Focus on what your child does right not on what he or she gets wrong.
- Reinforce patterns of success. Show how proud you are of your child's achievements, however small.

> **You've Got What It Takes**
>
> You know your child better than anyone. So that makes you your child's best talent scout. If you think about it, you have a warehouse full of your kid's success stories. Be prepared with those stories. They will help you focus on those qualities she or he can use to succeed.

Section 2
Winning Strategies

Chapter 6 — How to Start Coaching and When

In October Maleena was passing math, but just barely. As the year wore on, her grades slipped slowly. In April, when her dad offered to help, she snapped at him. "There's no point. It's too late to pass now. I'll just have to repeat math next year."

What Maleena and her dad didn't know was that repeating a course is no guarantee of success if there's no change in strategy. The beginning of a new school year is the best time to begin your new job as math coach. But it's never too late to lay the groundwork, build the foundation, and develop the homework habits needed for eventual success.

Preseason Warm-Up

As soon as you realize that your child needs help, make an appointment with the math teacher. But never try to do this in the first week of school. Why? Because by the second week your kid may have a new teacher! Ask the teacher for a course outline or a list of the topics to be covered during the school year. Explain that you are not trying to undermine the teacher, but just trying to support your kid's efforts to succeed in math. This lets the teacher know that you are a motivated participant in your child's progress.

It's not unusual if the teacher can't give you an outline right away. This doesn't mean that she or he hasn't thought about the course. It only means no extra copies have been made yet. Suggest that the teacher send the information home with your child when it's available. Don't be surprised if the first topic of the year is "review." As your kid's coach, this "preseason" review provides you with two important opportunities. First, it allows you to see your child's weaknesses while there's still time to set up a training schedule. Second, it gives you time to get the supplies you will need for a successful season.

> For other materials that will help get you started and keep you going, look on page 62. Then try page 63 for descriptions of the topic books available as part of the Yes You Can! program.

Stock Up on Supplies

Here's what we suggest to get you off and running.

Get your hands on a math textbook for your kid's grade level. *Any* textbook will do. New textbooks are very expensive, but you can often find usable math texts in secondhand bookstores, flea markets, and library sales. No matter how old or tattered the book is, you'll be able to find most topics that are on the course outline. Having a textbook at home is important even if your kid tells you: "We don't use a textbook."

Go to the library or a bookstore and look for parent-oriented materials.

Find materials written for earlier grades. They can help bring you up to speed. You'll be able to see what topics your child is expected to understand already. They can also help you diagnose difficulties and think up ways to correct them later on.

Once you've got your books, don't be afraid to open them! You don't have to know everything that's inside. That's not your job. You're the coach. These materials are only coaching manuals. Chapter 14 will give you some tips on how to use your math books and material effectively.

Coaching in the Real World

The ideal way to coach math would be like the ideal way to build a high-rise:
- **Prepare the worksite.** "Clear out" bad habits, negative attitudes, misconceptions, and weaknesses.
- **Establish a foundation.** Build good habits, reinforce positive attitudes, and strengthen basic skills.
- **Erect the upper stories.** Work on one layer at a time by progressing through an orderly sequence of schoolwork, assignments, and homework.

In the real world, this ideal order isn't attainable. Short of taking your child out of school, there's not much you can do to put the upper stories—homework—on hold while you focus exclusively on clearing the worksite and reinforcing the foundation. You will have to balance conflicting demands.

We won't pretend this balancing act is easy. Whether you start in September or "midseason," it's tough for you and your kid. But unlike some school coaches, *you* have a continuing contract with your child. You'll be around next season when your efforts start to pay off. And they will! So keep your eyes on that prize as you maneuver around your four main coaching challenges:

1. **Get the homework done.** Every weekday your child has to work on the upper stories at school and complete assignments at home.
2. **Overcome learner resistance.** The day-to-day buildup of math ideas at school is relentless. So your child may resist anything not obviously related to homework with a well-known cry: "We did that stuff *last* year."
3. **Choose foundation topics.** Focus on the fundamentals that will do the most to support the math being taught at school right now. Ask yourself: "What will maintain current skills so they don't get lost?"
4. **Find time for what's important.** As an adult, you know that fundamentals are important—they will pay off with long-term rewards. But fundamentals are not as urgent as homework. So it's hard to find time to focus on them.

> **Urgent or Important?**
>
> *Important* things contribute to long-term goals but do not demand immediate action. *Urgent* things have deadlines and demand immediate action. But what's urgent (a ringing phone) may not be important (wrong number!). And what's important (making a will) may not be urgent (I'm not even sick!). Math homework is both urgent and important. It has to get done every day, regardless. As math coach, you need to know how to handle homework right away. Chapter 7 offers strategies for dealing with homework. Chapter 8 discusses learner resistance and foundation topics. And Chapter 9 suggests ways to carve out time for fundamentals.

Chapter 7

Handling Homework—Keeping Up the Daily Grind without Getting Ground Down

Back in the 40s and 50s, it was common for a woman in labor to be left alone in a sanitized hospital room with only a bell for company. We now know that women left alone feel a lot more pain than women who have the company of a birthing coach. The coach does not have to be a doctor, a nurse, or a midwife. A coach—whether friend, spouse, partner, or even a volunteer—makes the whole process more bearable.

The Warm Body Hypothesis

Here's a widely held view of the parent's role in homework: provide a quiet, private room so the child can work alone behind a closed door. That pattern may benefit some kids, but it's not likely to help the child who is struggling in math. We have a different view—the warm body hypothesis: *A learner's chances of meeting math success are greatly increased if another warm body is present.*

We are not revolutionaries. Current research has tracked the school performance of two groups of youngsters. Both groups came from inner-city environments, crowded homes, and large, poor families. Group 1 children were born in the United States to English-speaking parents. Group 2 kids were recent immigrants. English was their second language and not spoken at home.

Family size, income, living space, and environment are powerful predictors of school performance. So neither group was expected to perform well. But Group 2 children had an extra disadvantage—being taught in a new language. You'd expect their performance to lag, right? But the outcome was just the opposite. The immigrant group performed significantly better than the nonimmigrant group.

Why? Researchers concluded that a "warm body" made the difference. In the immigrant group, the kitchen became the family learning center every evening after supper. All of the children—regardless of age—gathered around the table to do their homework. At least one parent remained in the room, even though he or she could not speak English.

Setting Up the Coaching Center

Setting up your coaching center is where you can apply the learning strategies we present in this book. You don't need a private room, but you do need a table with space for your math materials and a clear working surface at a kid-friendly height. Choose a corner or curve where you can sit beside or at right angles to your child. Don't (if you can possibly help it) sit across from each other.

Your coaching center doesn't have to be completely quiet. Dishwasher sloshing is fine. Low music on the radio may be okay. But the TV has to be OFF. Sorry, no exceptions. Other family members can read or do quiet chores. Very young children can do their own "homework" with their own "stuff" (such as Lego) at the other end of the table. But they must understand that the coach is available only to the math student.

The Role of the Math Coach during Homework

Rule 1: Parents should coach, not rescue. So you don't—*you mustn't*—do the homework. Providing right answers for your child to carry to school is not a coach's job. It's far better for your kid to go back to class with a question about the math he or she doesn't understand. Any reasonable teacher should accept this as a sincere attempt to complete the homework. (Chapters 12 and 13 show how to "frame" such a question for the teacher.)

Rule 2: As coach, you are on duty at all times during the session. Other family members can read, draw, or build with Lego, but not you. A coach must be completely attentive to the math student, alert to everything the child says or does, ready to coax, encourage, and support as needed. The advantage of this close scrutiny is that you can see the split second your child stops paying attention.

Rule 3: Don't scold. Kids who are struggling in math often "blank out" or daydream. This is a perfectly natural response that helps them deal with that awful hopeless feeling. In class, the teacher usually can't detect it soon enough. At home, you can. When you do, don't judge. Gently bring the learner back to the task at hand. In a matter-of fact tone say something like: "I can see your attention has slipped. Can you show me where you left off?" Be courteous—you are treading on tender ground here.

Rule 4: The coaching situation requires very intense interaction and is often draining for coach and child. That's why it's so important to set realistic time targets. A kid whose past includes a lot of time staring at math in hopeless frustration likely isn't used to focusing fully on math. So you need to build in rewards for paying full attention during homework. If either of you has stopped feeling attentive, it's time for a break.

> **You're More than a Warm Body**
>
> If a parent can boost grades simply by being present while kids do their homework (in another language!) just think what *you* can do! Even if you aren't a math teacher; even if you don't remember much math, didn't do well in math, or didn't graduate from high school—if you can read this page, you can be much more than a warm body. You can be your kid's booster, interpreter, and coach.

Chapter 8: Focusing on Fundamentals

No homework tonight? Wonderful! Perhaps the two of you can enjoy a reward. But first, be sure it's true. Learner resistance takes many forms, and kids who are struggling often fool themselves about homework. "I can do the rest on the bus." Don't scold, just calmly bring your child back to reality.

The Challenge of Learner Resistance

For most strugglers, reality is that even getting all of their homework done is not enough. Gaps from the past must be filled in; foundation topics must be maintained. But strugglers often resist working on fundamentals. Let's look at three sources of learner resistance and ways to overcome them.

Resentment

As coach, you need to be aware that foundation work could seem like extra duty—or even punishment—to kids who are already trying hard. You can't always know how much it's costing your child to make a seemingly small effort. So it's vitally important to develop a reward system that recognizes the importance of foundation work. (See Chapter 16 for suggestions.)

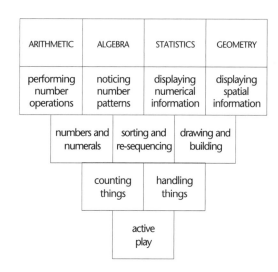

Relevance

"When am I ever going to use this stuff?" This line is a common sign of learner resistance. Here are two possible—and equally important— answers: (1) Your child's future is ten or twenty years ahead. We can't know for sure what may be important then. (Who knew twenty years ago that today's grade 3 students would be using e-mail?) But we do know that the thinking skills developed in math will never go to waste. (2) Math is a prerequisite for many careers and courses that don't involve doing much math. Employers and admissions officers use math as a "gatekeeper." They reason that those with the personal discipline to get through the "math gate" are likely to succeed in a new job or course.

Embarrassment

Often the key to a problem in math lies in a small misconception left over from an earlier grade. Many kids who are struggling fear that revisiting the work of earlier grades makes them look stupid. You may hear lines like: "But I passed grade 3 math already." When this happens, compare math to an inverted pyramid. The big ideas at the top rest on a narrow base. If the small ideas at the bottom are not firmly mortared in, the whole structure can crumble.

Choosing a Suitable Foundation Topic

The math coach must be ready to adapt to changing conditions on a daily basis. Some days it's best to focus on the foundation topic that would do the most to support the homework or prop your kid up in class tomorrow. Other days it's best to work on filling in gaps from the past, or doing maintenance—sharpening skills that have not been used in a while.

Gaps from the Past

You may have to go back more than one grade level to locate the original misunderstanding. Chapter 14 discusses how you can use textbooks to trace the development of math concepts. You might think that official curriculum documents could help. But these are often huge and hard for a layperson to interpret. Page 63 describes how you can get a predigested, all-grade chart. For now, it's useful to know that most modern math courses are woven from four main strands: Numbers, Statistics, Shapes, and Patterns.

Curriculum Strands

Numbers: Number Sense	Recognize how numbers link to objects and numerals. Use number operations and properties in computation and estimation. (Arithmetic)
Statistics: Data Sense	Collect, organize, analyze interpret, and display data. Use data to predict the likelihood of an event. (Statistics and probability)
Shapes: Spatial Sense	Draw, construct, compare, classify, and relate geometric shapes (1-D lines, 2-D figures, and 3-D objects). Use appropriate methods to solve measurement problems. (Geometry and measurement)
Patterns: Pattern Sense	Analyze and describe mathematical patterns and logical relationships using tables, equations and graphs. (Algebra)

Math Maintenance: Flossing the Brain

All parents know this truth: one hour of flossing after a month of neglect won't prevent cavities. Teeth require regular maintenance. By flossing your own teeth, you set an example for your child. The same is true for math. By embedding math thinking into daily life you can set another powerful example. This strategy will reduce your kid's discomfort with math ideas, and make using them feel as natural as walking and talking.

Chapter 9   Teaching Your Kid to Walk and Talk Math

By using the ideas in this chapter—adjusted for the age and stage of your child—you can create a math-rich environment at home and open up time for fundamentals. As you read, you'll begin to recognize that daily life is full of opportunities for injecting "math moments."

Noticing Numbers

To help your child associate numbers with something "real" (food) and meaningful (money), go shopping together:

- *In the aisles:* Have your kid look at shelf tabs and read out the total price and unit price for each item. Together, use these numbers to help make buying decisions.
- *At the checkout:* You unload the groceries as your child watches the cashier's display screen. Ask: "Do the prices look right?" Let your kid take charge of the cash register tape.
- *At home:* Get your child to check each item against the cash register tape as you put the groceries away. Ask: "How much did we pay?" "For how much food?" "What percent in tax did we pay?" "What did it add to our total?"

Exploring Shapes and Spaces

To help your child associate words from books (e.g., symmetry) with real things (e.g., butterflies), have your kid:

- *Choose a 2-D shape:* For example, find triangles in nature (leaves and petals) and triangles in built things (yellow yield signs). Sketch a triangle, then trace its perimeter (outside edge) in purple. Cover its area (surface) in red. With eyes closed, describe its shape, perimeter, and area.
- *Choose a 3-D object:* For example, choose a cube and use the 2-D ideas above to explore it. Roll a sugar cube in his or her fingers. Trace its edges. Look at its top, bottom, and sides. Color its entire surface area. Figure out what part of the cube is its volume (the space inside the cube).
- *Find out how things for sale are measured:* Find out how towels, building lots, gasoline, carpets, and concrete are sold. By volume (quarts, liters, cubic yards)? By area (square inches, square meters)? By length, width, height, or depth (yards, inches, centimeters)?
- *Explore spatial relationships:* Find symmetry in nature (butterflies have it; clouds don't) and symmetry in built things (keyholes have it; keys don't).

> **Explore More**
>
> From a textbook, choose more 2-D shapes (e.g., circles, squares) and 3-D objects (e.g., spheres, rectangular solids). Explore a new one each week.

Surveying Statistics

Encourage your child to notice statistics on TV, radio, and in newspapers or magazines. Try asking these questions:
- *Commercial ads:* Can you find an example of a product survey (e.g., 4 out of 5 dentists agree)?
- *Sports news:* Can you state game or player statistics (e.g., averages)?
- *Lotteries, horse races:* What are "odds" (e.g. 5 to 2)?
- *Elections:* What are "polls," "sample size," and "uncertainty"?
- *Business news:* Why use graphs and charts to display data?
- *Weather reports:* What is "mean" monthly temperature?
- *Weather forecasts:* What is "probability" of precipitation?

Recognizing Patterns

Math often uses *symbols* to represent numbers and describe *patterns*. To help your child grasp the true meaning of those words, have her or him explore calendar patterns.
- *Play with the numbers:* This picture is a 2 × 2 block from an ordinary calendar page. Have your kid add along each diagonal. Ask: "What do you notice?"

13	14
20	21

13	14
20	21

- *Repeat the experiment:* Ask your child to repeat the number play with several 2 × 2 blocks on real calendar pages from different months. Ask: "What do you notice this time?" "Does the same thing happen?"
- *Recognize a pattern:* Any result that happens repeatedly can be called a pattern. Ask: "Have you found a calendar pattern?"
- *Describe a pattern in words:* Explain one way to describe the pattern is: "Top left plus bottom right is the same as bottom left plus top right." Then explain that the pattern could be written in another more useful form:

Top Left	plus	Bottom Right	is the same as	Bottom Left	plus	Top Right
TL	+	BR	=	BL	+	TR

TL + BR = BL + TR describes the calendar pattern with symbols instead of words and numbers. Using symbols to represent numbers and describe patterns is what algebra is all about. There are many more patterns hidden in a calendar week or month. Get a calendar big enough to write on. Encourage your child to build up a day-by-day pattern collection using symbols to describe them. For example:

Sunday + 1 = Monday, so Su + 1 = M
Thursday = Friday − 1, so Th = F − 1

Chapter 10

Plugging into Math—How to Get the Current Flowing

We've all had this experience: you plug in the toaster and it doesn't work. Why? Because a fuse has blown or a breaker has flipped. What do you do? If you have the time and are electrically minded you grab a flashlight and head off to investigate the electrical panel. If you're running late or dread electricity, you plug the toaster into another outlet. Both of these actions have the same result—they get the current flowing again so you can get on with your day. In this chapter we will discuss some strategies you can use to help your kid get plugged into math.

When the Learning Circuit Gets Blocked

Because of the way math is generally taught in schools, your kid may have developed only one way—one circuit—to learn math.

In a typical math class, the teacher sends information through "mini" lectures and students receive that information by listening and seeing what the teacher writes on the board. For some kids this learning circuit is always open. They rarely have trouble understanding what is being said or shown them—they seem to absorb math symbols and concepts like sponges. But when this math learning circuit gets blocked, kids can't make sense of the information they receive. They get frustrated and angry with themselves for not "getting it" or—worse—for being "dumb."

The real problem is that kids don't have the experience to recognize that there are other ways to get the current flowing again. They have been conditioned to think that there's only one way to learn math. That's where you come in. And here are some strategies you can use to help your child activate new learning circuits.

Recognize Your Kid's Learning Style

Think back on your child's development from the time she or he was a baby. This should give you some clue about the kind of learning style that works best for your kid. Look at the learning styles listed in the chart on the next page. Do you recognize any of them?

Your child has probably developed a combination of learning styles. And all of them can be used to help open up new math learning circuits. Think about how you might take advantage of your kid's learning style and try to figure out ways to get around the limitations that style might impose. Remember that the major problem your child may have is not that he or she has no learning style, but that no circuit has been opened up to use that style effectively.

The Builder
- has a highly developed sense of touch and is always making things
- interprets the world three dimensionally
- seeks concrete understanding of his/her environment through hands-on involvement

The Artist
- is always doodling
- represents experience visually through painting and drawing
- responds enthusiastically to color, shape, and texture

The Talker
- is a chatterbox always wanting to tell about what she/he has learned
- may have wide-ranging, but fragmented, knowledge
- may be forgetful

The Writer
- is always writing poems, stories, journals
- writes to make sense of the world and to make experiences concrete

The Teacher
- "instructs" others about what he/she has learned
- needs a structured setting to relate to knowledge
- has a good memory

The Watcher
- is shy about self-expression, but extremely attentive to everything going on around him/her
- must "see" everything before being confident enough to relate knowledge

The Reader
- reads voraciously, anytime, anywhere
- makes concrete associations with written, not spoken, words
- has a highly developed inner voice and active imagination

Engaging the Senses

Touch
- Read a math question aloud while touching each part as it's spoken.
- Keep one finger on the question while writing out the solution with the other hand.

Sight
- Make a color key for math operations. For example, highlight all multiplication symbols in blue.
- Use colored "sticky" notes to translate instructions. For example, write "Divide" on a pink note and stick it close to a division question.

Hearing
- Listen with closed eyes while a math question is read aloud.
- Speak into a tape recorder while working out a problem, then play back the results.
- Memorize important math "facts" using rhyme or song.

Get the Current Flowing

Once you have determined your kid's learning style, what can you do to activate new learning circuits? The most efficient method we know is *active learning*, which is a way of physically engaging your child's senses in the learning process. Using the senses to open up new learning circuits may give your child options for learning math he or she has never had access to before. In the box are some ideas we think work to get the current flowing. But be patient, it may take some time to find the best strategy or combination of strategies for your kid. Your kid is the expert here. More than likely he or she will be able to "feel" if something works before being able to tell you why.

Chapter 11

Making Sense of Math—The Power of Reading

I thought I was doing better in math. But I just flunked the last drill I did and now I'm not so sure. I thought I was doing the right thing. I looked at the page and saw a whole bunch of two-line questions. There were no sign things. But the tops were all bigger than the bottoms, so I subtracted. I even finished early. But then my teacher told me that I should have multiplied, not subtracted. Why didn't she say that sooner? — Cecilia

If All Else Fails—Read the Instructions!

Reading—making sense of words on a page—is as important to mathematics as it is to any other subject. This astonishingly simple idea may never have been communicated to your child during his or her school years. Many people—and kids in particular—DO NOT READ INSTRUCTIONS. All of us have tried to assemble something without reading the instructions. We were in a rush, we were impatient, the guests were arriving any minute. So we took a *shortcut*.

Shortcuts in reading instructions are fatal to math understanding. And the one place they occur with alarming frequency is in doing drill sheets. Consider Cecilia's experience: She did not read the three-word instruction at the top of her drill sheet—MULTIPLY THE FOLLOWING. Because she was anxious to do well and finish on time, she took a shortcut and ended up with a page full of wrong answers and a heart full of frustration. Cecilia, like millions of kids in school today, didn't realize that she *had* to read the instructions if she wanted to make sure she did the right thing. Breaking the pattern of not reading instructions is a major step toward greater math understanding.

Don't Miss the Reading Boat

Suppose this morning you read a cookbook, a newspaper, and an insurance policy. As an adult, you know enough to bring a different mindset to each one. Many kids don't. Most of what kids read in the early grades of school is based on fiction. They do not read much material based on explanation or instruction. Since kids don't understand the structure of this kind of writing, they may take the easy way out—by passing their eyes over the page not really *reading* the words. Because kids don't know how to interpret the words, they often attach no meaning to what they read, especially if it is about math. Kids who are having difficulties in math associate reading with word or story problems they cannot solve. They soon convince themselves that there is always something tricky about the words being used and they will never be able to figure them out. They don't expect to *comprehend*.

Developing Comprehension Skills

Reading with comprehension is a skill that can be learned. This chart shows the characteristics of unskilled readers and presents some strategies for developing effective reading skills. Do you see how these improved skills can open new learning circuits for kids who are struggling in math?

Unskilled Readers ...	Skilled Readers ...
Don't realize that textbooks are organized in a structured way.	Scan the textbook to find main headings; use the headings to develop a chart to fill in as they read.
Start reading without preparation.	Think about what they already know about a topic before starting to read.
Start reading without knowing why.	Record their reasons for reading.
Are easily distracted because they don't expect to understand what they are reading.	Are focused because they expect to understand what they read; they pause frequently to check their understanding.
Don't know what to do when they don't "get it."	Ask questions if they don't "get it."
Stop thinking after they read so they don't know what was important.	Continue to think after reading; reflect on major ideas; write down what was important.
"Tack on" new information; they don't incorporate it into what they already know.	Incorporate new information into what they already know.

Adapted and modified from *Strategic Learning in the Content Area*s, Wisconsin Department of Public Instruction

Reading Strategies to Help Your Kid Make Sense of Math

Here are some strategies your kid can use to develop effective reading habits in math.
- Don't jump ahead.
- Cover the drill sheet or exercise with a blank piece of paper. Slide the paper down until the instructions appear.
- Highlight the main instruction or record it in a notebook.
- Read an instruction aloud and explain what it means in your own words. Answer these questions: "What math 'action' must be taken to follow the instruction?" "What will the answer look like?"
- Read an instruction. Then cover up the sentence and mentally picture what you are supposed to do (add, subtract, divide, multiply, etc.).

Math books are not like fiction; they are meant to be read *slowly*. And effective reading skills are critical to understanding math. One way these skills can be reinforced is by talking out loud. Chapter 12 discusses strategies you can use to get your kid talking about math.

Chapter 12

Getting Your Kid to Talk—The Role of Verbalization in Learning Math

I have taught both children and adults. Teaching adults is so much easier because they can—and do—tell me exactly where they stop understanding. With kids you never know. They're not very verbal with "authority" figures and they're usually embarrassed to say "I don't get it" in front of their friends. Do all those bright eyes mean they're following the lesson? Or is it just incomprehension shining back?

— Ruth Katz, Grade 7 math teacher

How Talking Helps

Learning specialists everywhere agree that talking plays a vital role in the learning process. Children don't get enough opportunity in school to talk out loud to an adult who is really listening. Your kid's teacher has to tune in to 30 voices. How much time do you think he or she has to spend listening to your child? To help, many schools have introduced cooperative learning strategies. Cooperative learning situations encourage children in groups to problem-solve by talking to each other. Talking to other children can be enormously valuable.

But it's not the same as talking to an adult who cares about how a child progresses in math. That's where you come in.

A walk down the hall of any school would confirm that students do a lot of talking. They also spend a lot of time in school listening—to their teachers and to other kids. Active listening is an important pathway to learning. But listening can't engage a child's brain in the same way talking does. The great thing about talking out loud is that it's really thinking out loud. Verbalizing engages a kid's mind in a way nothing else can.

Talking out loud not only communicates thoughts to a listener; it also communicates thoughts to the self, using a different part of the brain. Verbalizing lets your kid hear his own voice, hear herself think differently than when thoughts are formed silently. Talking launches ideas and reveals uncertainties and gaps in logic. Feedback reinforces the problem-solving process and keeps "the talker" on track. And feedback is something you can provide without being a math guru or a teaching expert.

What You Can Do

Talking is how you access what is going on in your kid's head. This is something a teacher seldom has time to do. Even when a teacher can devote time to listening, kids may be reluctant to display any uncertainty in front of their ever-critical peers. Risk-taking in the classroom isn't often understood or rewarded. But you can reward it simply by paying attention to what your child is saying. By encouraging your kid to tell you everything he or she is thinking while working through a math problem, you can find out what is going on in your child's mind. This can be a real eye opener! And you don't have to be a mathematician or even a high school graduate to recognize where your kid is going off the rails.

Strategies for Getting Your Kid to Talk Math

Verbal mediation

Verbal mediation is educational jargon for a very simple idea. It combines *verbalizing*—talking out loud—with *mediation*—controlled interruption. This strategy means that you do most of the listening and your kid does most of the talking. Your coaching goal is to make your child feel comfortable with the idea of thinking out loud in the presence of someone who will give his or her thoughts a spin and bounce them back.

Tell me more

This strategy works hand in hand with verbal mediation. In fact, it's a good way to constructively interrupt but keep the process moving along. Your coaching objective is to keep your child talking and on target. Try nudging your kid with the occasional question or comment such as: "Tell me more." "It's not clear to me how you got from here to there." "Could you run that by me again?"

Frame the question

Once the nature of a problem is revealed, your kid can often get past it. If the problem is beyond your kid's current abilities—and yours, too—this is where you can make it clear that the teacher has an important role. Try saying: "I'm only the coach, I'm not the expert on this. How can we make up a question for the teacher that will get us the information we need?" Remember you don't have to solve the problem, but you have enough life experience to give your kid the tools to frame an appropriate question. Once the question has been framed verbally, carve it in stone by turning the talking into writing. (See Chapter 13 for the importance of framing written questions for the teacher.)

Chapter 13

The "Write Stuff"—The Role of Writing in Learning Math

Just as breathing exercises help integrate body and mind, writing ... distills, crystallizes, and clarifies thought and helps break the whole into parts. — Stephen R. Covey, "The 7 Habits of Highly Effective People"

New educational research urges teachers to have kids write about the math they are learning and the ideas that puzzle them. Unfortunately this strategy takes time—that oh-so-scarce classroom commodity—and "covering the curriculum" usually comes first. But we think writing is a wise way to spend coaching time.

Plug into the "Write" Circuit

For many learners, the muscular act of putting pen to paper seems to engage the brain in three ways: (1) The brain signals nerves and muscles to begin the writing act. (2) The brain chooses the words and reasons out the order in which they are to be written. (3) Seeing those words in the chosen order can stimulate the brain to plug into a problem in a new way.

Write to Think More Clearly

Many kids are lazy thinkers and the gaps in their thinking show in their writing. For example, this type of question shows up frequently on math tests.
Question: A car traveled 100 miles in 2 hours. What was its average speed?
Franklin's answer: The car's average speed was 50 miles per hour.
Zachary's answer: 50.

Technique	How to Use It	How It Helps
Write what you say.	Record ideas, solutions, or problems as you talk about them.	Captures ideas that might get lost, encourages focus, and improves memory.
Write what you think.	Try to sort out meaningful words from fuzzier words.	Brings your thinking to light so you can see the separate parts and look for a pattern.
Write in complete sentences.	Try to distill your thoughts and comments into one or two sentences.	Pulls parts of your thinking together so you can "get" the whole picture or pattern.
Read your writing aloud.	Ask yourself: Does it make sense? Are there better words to make your idea clearer?	Helps you decide if what you wrote will be meaningful to others (e.g., your teacher).
Compare your written answer to the original question.	Ask yourself: Did you do what your teacher asked? Did you do everything the teacher asked?	Improves your marks; helps you become a more effective learner.

When the test is handed back, Franklin gets full marks and Zachary gets zero. That's because the question asked for speed and Zach left out the units—miles per hour (mph). The 50 shows that Zach knew how to divide 100 by 2, but without any units the 50 does not represent speed. Zach likely knows that the speed would come out in miles per hour—"Isn't it obvious?"—but the teacher can't mark what's still in a student's mind.

Writing lets you, the coach, detect this kind of gap and analyze your child's thinking. Then you can apply adult common sense to childish misconceptions and omissions. This is one of the strengths you bring to the coach-learner partnership.

Write to Test Understanding

Many teachers and students think of math as something to do, not as something to write about. Typical textbooks introduce new math with minimal language. They provide one or two worked examples and then ask kids to do dozens more on their own. For learners who thrive on symbols, this approach is effective. For those who struggle with math, it's not. The chart on page 36 describes some writing techniques your child can use to improve math understanding.

Write to Communicate with the Teacher

Dinner's over. You ask your child to clean up the kitchen. Ten minutes later you hear the door slam. Your kid truly believes the job's done, but you find the counter unwiped and three dirty pots on the stove. It's a common scenario, right? But what if you got your child to write out a job description first? Then you'd know he or she understood what a complete job requires. The writing would be a communication pipeline.

Writing can be a communication pipeline between home and math class, too. But your child has to do it, not you. One of the most important things you can do as a coach is to get your kid to record and bring home any notes the teacher gives in class. This isn't always easy. Students may not understand everything the teacher writes down. Or they may miscopy notes off the board. But you can use a textbook and your adult judgment to figure out what's missing.

The more notes your child records and the more the two of you work together to understand those notes, the greater your kid's confidence will be that the two of you are on the same side. Your kid's commitment will strengthen and note-taking will improve. (One thing to remember—make sure your child's eyes are okay. Especially during puberty, vision can change dramatically.)

> **Keeping the Communication Pipeline Open**
>
> In Chapter 12 we emphasized the importance of "framing the question" for the teacher in spoken words. Writing the question down puts it into a form the teacher can understand and strengthens the link between home and school.
>
> Another way to keep the communication pipeline open between home and school is to have your child record questions and the teacher's answers in some systematic way. Modern learning experts recommend keeping a learning log. Go to Chapter 15 to find out how learning logs can be effective tools in learning math.

Chapter 14

How to Use a Textbook as a Coaching Manual

Since Chapter 6 we've been urging you to use a textbook as a coaching manual. But we should come clean about the textbook's limitations. Many don't incorporate what is known about how kids learn and don't do much instruction. Publishers create the books that way because most math teachers prefer to use their own teaching methods. They want a textbook to provide lots of exercises.

Despite its limitations, your at-home textbook can be used as a coaching manual for both daily homework and at-home work on fundamentals. But to use it effectively you should analyze—break down and identify—the book's structure. First look for scope and sequence. Then look for instructional design.

Scope and Sequence

"Scope and sequence" is educational jargon for "topics and order." It's helpful to make a chart linking the course outline you got from the teacher to the scope and sequence of your book. Use the Table of Contents to get an overall picture of what's available in your at-home textbook. Make a photocopy of the course outline, mark it up with page and chapter numbers from your at-home book. Most modern math texts have an index, so you can (fairly) easily look up a particular topic. Always look at the earliest page reference first. Keep looking until you come to an explanation or a definition.

Instructional Design

"Instructional design" is educational jargon for the structure or pattern a book uses to introduce, develop, and practice new ideas. Most math textbooks use the same general structure for every topic. Arithmetic, statistics, geometry, algebra—all get the same treatment. The chart on the next page shows a very common pattern. Compare it to the structure of your at-home textbook. What special limitations does the structure of your book impose? How does it fit your child's learning style? (If the school has issued your kid with a textbook, it's worthwhile analyzing it as described above.)

What's in a Typical Math Book	How It Affects Your Kid
Topic or skill to be learned is introduced in isolation.	No discussion of how the topic is linked to math topics covered in the book or in earlier grades.
One or two short sentences introduce new words.	The ideas are too condensed for most students. There aren't enough real-life links.
Diagrams, photos, or graphs illustrate the topic.	Captions and labels fail to convey the link between pictures and sentences.
One or two sample questions are given with sample solutions.	The reasoning behind the solution is not explained in enough detail.
Basic exercises have some "simple" questions to let students practice the new skill or topic in class.	Basic exercises may be poorly organized. Questions do not progress from easiest to hardest.
Advanced exercises have more difficult questions. These are usually assigned for homework.	Math whizzes get most of these done in class. Strugglers do them at home, with no teacher to help.

Foundation Coaching Tips

Your at-home textbook is a great source of practice drills for foundation work. These drills should be done in your child's learning log. But don't choose questions for this purpose until you check the back of the book. There'll be a section called something like "Answers to Exercises." Examine it carefully. Does it give answers to *all* questions? Many texts give answers to selected questions. You'll want to know which ones before you select a drill for your kid.

Every textbook ever printed contains some "typos." In math books that means some of the answers in the back of the book are wrong, especially in the first printing. Wrong answers can cause enormous frustration. (If after reasonable effort—say three tries—you can't get the answer that's in the back—stop. The answer may be wrong. Go on to the next question, or go back one.)

Plan for success—avoid "basic exercise" blues. Don't automatically ask your kid to do question 1 first. Examine sample solutions carefully. Then look closely at the basic exercise. Give your child a taste of success by asking him or her to do the question that looks *most* like the ones in the sample solutions.

Homework Coaching Tips

Most math ideas can be approached from many directions. Your at-home math text can give your child more chances to succeed:

- If your kid didn't understand the teacher's explanation of the homework topic, the explanation in your at-home textbook may help.
- If you have a textbook for an earlier grade, look in its index for the same topic. Many topics appear several grades in a row. The earlier the grade, the more basic the explanation will be for your kid and for you.
- If your child has a school-issued textbook, its treatment of the topic may provide clarification. Often the teacher will have used a different approach.
- As coach, help your kid compare how topics are presented in different books. Which makes more sense to her or him? This may help your child frame a question for the teacher showing exactly what she or he doesn't understand.

Chapter 15

How to Develop Helpful Habits and Routines

> *"Math class is tough..."* —Talking Barbie before Mattel changed her tape in response to cries of outrage.
>
> Probably more "negative" tapes have been created about math than any other subject. But like Mattel you can help change your child's tape and help her or him develop a more positive attitude. Modern learning experts say that the single most helpful routine kids can develop—for math or any other subject—is the learning log habit.

What Is a Learning Log?

A learning log is *not* the same as a math notebook. A math notebook is organized according to the teacher's expectations. It's used at school to record notes from the board or overhead projector. And it's used at school and at home to do assignments.

A learning log has a different function. It's a journal or diary for recording the learner's *personal* comments, ideas, and framed questions. A learning log can be organized in any way that works for the user—your kid.

What Does a Learning Log Look Like?

A learning log has to be portable and made of paper. A computer—no matter how portable—is not practical for a math journal. A learning log could be a collection of index cards wrapped with an elastic band. Or, if the teacher agrees, the log could be kept in a special section of your child's regular math notebook. But for most kids, a separate diary such as a hard-cover journal works best. Kids will take it more seriously.

What Goes into a Learning Log?

Anything the user likes, really. But here's what we suggest.
Discoveries: Did I learn something new today?
Puzzles: Did something stump me?
Framed questions: Here's what I'll ask the teacher.
Solutions: Here's what the teacher answered or explained.
Successes: Did I answer a question in class? Complete all the homework? Get half of it right?

Can you begin to see the power of writing in a learning log? This well-tested strategy can help your child get control. And with you as coach, you'll be well on your way to achieving the goals the two of you have set for math success.

Developing an Effective Drill Routine

Developing systematic work habits—a routine—can set your child up for success in any subject. Since drills can be a useful study aid to prepare for tests, here are some strategies your kid can use to develop an effective drill routine.

Before starting the drill

The following routine is really effective, but most kids aren't used to doing it, so you will have to persist. But persisting will be worth your while.
- **Date** the top of every working page.
- Give the drill a **title** that includes key words to identify the kind of math to be done such as *Adding Fractions*. These words are likely to show up on a math test.
- Record the **page number** and **question numbers** of the drill.

While doing the drill

A surprising number of strugglers try to use the question number as part of the exercise. Developing the following habits will help your child focus on the question itself.
- Use the pointer finger of your nonwriting hand to locate the question number.
- Record the question number on the working page.
- Then use your pointer finger to cover up the number in the textbook.

After the drill
- Highlight the questions your child gets right on the first try in a color, such as blue, that means "good" to him or her.
- Highlight the questions that need a second try in another color, such as yellow. These are the questions to focus on when preparing for a test.

Other Helpful Habits
- Look for math around the house.
- Browse flyers and newspapers ads. Find out how towels are measured and sold.
- Help your child associate measuring units, such as pounds and feet or meters and kilograms, with materials sold that way.

Some Good Habits for You

You are your child's best chance for developing positive attitudes toward math. Here are some guidelines to keep you on the right track.

Avoid frustrating your child. Recognize the source of your own frustration. You may be angry that the system let you down and is now letting your kid down. Many educators are frustrated, too. After seven generations of new math curricula, problems with math are still widespread. So they're no reflection on you, your parenting skills, or your child's intelligence.

Clean up your language. Don't call yourself a math dummy. Don't call anyone a nerd. Speak up when songs, TV, or relatives make unkind jokes about math ability.

Keep your sense of humor!

Chapter 16   How to Recognize and Reward Success

The bank Gary works for has launched a school outreach program. Bank employees go into the schools and talk with kids about career planning. Here's what he heard two minutes into his first interview:
GARY: "How are you doing in math?"
STUDENT: "Not very well. I'm not good at it."
And two minutes later:
GARY: "What do you see in your future?"
STUDENT: "I want to be an astronaut."
Gary was astonished. But the student seemed blissfully unaware that these two statements contradicted each other.

Why It's Hard to Motivate Modern Kids

It's very common for kids to carry a negative image of their present-day selves alongside an unrealistically positive image of their future selves—and see no contradiction. Many believe, regardless of their current living circumstances, that they will have the "good life" they see on TV—upscale living quarters and seemingly endless leisure time. Most kids don't acknowledge that the people they see are playing fictional roles in a fictional world.

Students who do have career goals often name occupations they see on TV—doctor, lawyer, nurse, police officer. But TV seldom shows the effort needed to qualify for these jobs—it gives the illusion of *security* without effort. Because kids see no connection between school performance and future lifestyle, they can't be frightened into learning anything.

Psychiatrist William Glasser is the author of *Schools Without Failure*. He says modern kids are looking for "a role not a goal," but don't recognize that goals must be met on the way to getting the role.

The Value of Positive Feedback

Glasser uses the word "role" where most of us would say "identity." Your kid's identity may have been formed by years of negative feedback. "Bad" math marks send a very clear message: The work you do is not good, so you are not good, therefore I don't think much of you.

But it would sound—and be—false for you to tell your child he or she is "doing great" when nothing concrete that is praiseworthy has been done. You can create a different scenario by finding a way to get your child to do something—anything—so you can give positive feedback. Make it clear that *you* believe long-term goals are important, so important that you're prepared to reward every little step on the way. Make it worth your kid's while *now*.

The Value of Short-Term Targets and Rewards

You and your child have probably already set some long-term goals. "Get a C+ by June" might be an example. But if June is six months off, that's a long time for your kid to keep long-term goals in view. Your child lacks your ability to measure progress by setting appropriate short-term targets. So break the path to a long-term goal into very small, very realistic steps. For example, "Bring math notes home from school every day" is both small and realistic. Lavish praise every day your child does and don't complain the day he or she doesn't. This is easier if you develop a checklist that will detect the slightest sign of progress and measure it daily or weekly.

> " There is no failure here. We are here to learn... I will give you credit for what you learn towards that goal. If you learn a little, I will give you credit for it. If you learn a lot, I will give you credit for that."
> — William Glasser, MD

Check the box each time your child hits the target. (Use checkmarks, gold stars, stickers—any symbol that has meaning to your child.) Leave the box blank if he or she strikes out. This approach will give your kid credit for every little step. Any misstep will show up as a blank not a black mark. And your kid will experience positive feedback—success—long before math marks improve. Who judges when a blank box gets filled? In most cases, the coach will, especially at first. Make sure a few targets are worded so your child can self-evaluate.

Whatever form they take, the checkmarks are credits. They can be tallied up at the end of the week, and accumulated like frequent flyer points. You and your child can agree on a reward schedule. Turn in 50 for a movie; save up 500 for a pizza party.

Sample Short-Term Target Checklist

Target	M	T	W	T	F
Wrote in learning log about problem with today's class.					
Talked with coach about math idea seen on the way to school.					
Recorded page and question numbers of homework.					
Brought home everything needed for homework.					
Read instructions before starting homework.					
Tried all homework questions at least once.					
Recorded date and topic of next math test.					
Brought date and topic of test to coach's attention.					

Chapter 17: Tackling Test Fear—How to Coach Your Kid to a Personal Math Best

The coaches at an American basketball camp tried to find a more effective way to help their players improve their skills. After an initial assessment of each player's skill they divided the participants into three groups:
- *Group 1 practiced basketball skills 8 hours a day.*
- *Group 2 practiced 4 hours a day and watched training videos for 4 hours.*
- *Group 3 practiced 4 hours a day and learned positive imaging techniques for 4 hours.*

Which group improved the most? Group 3 did. Why? Because the players were able to visualize themselves getting through some tough plays. And they felt more in charge and calmer on the court.

The Power of Preparation

Three important strategies you can teach your child about preparing for a test in any subject are (1) learn, (2) review, and (3) practice.

Your kid's math notebook, learning log, and at-home math text can provide much of what your child needs to prepare for a test. The math notebook is a record of what has been taught (or not taught!) in the classroom, providing a "clue" about what might be on a test. Your child's learning log shows where she or he has struggled—areas to focus on before the test. And the at-home math textbook provides backup information and a review question source. The materials the two of you have developed during coaching sessions are also valuable test preparation tools.

The Power of Positive Imaging

Developing a positive attitude about tests is probably the most important thing your kid can do. Athletes, such as marathon runners, often use imaging techniques to help them prepare for a race. They visualize the whole race—from start to finish—and imagine how they will meet the challenges of the course, the weather, and their own fatigue. In short, they write their own scripts for success. They do what works for them, not against them.

Your child has probably accepted a "bad" script for math success. In many schools only "right" answers are rewarded, not the effort or the reasoning required to get *any* answer. Test failure results in low self-esteem and fear. Developing positive imaging skills is an effective way for your child to experience success, especially in tense situations like tests.

Rewrite the Script

Your child can rewrite his or her script and you can help. Ask your child to sit quietly and imagine taking a test from start to finish. (Many kids find this easier if they close their eyes.) Ask: "What do you feel like when you sit down at your desk?" "How many questions are on the test?" "Are there any you can answer?"

Kids who fear tests will probably answer that they feel "sick," that there are a "billion" questions on the test, and that they can't answer any of them. Your job is to prompt your child into rewriting his or her script. Each time your child comes up with a negative response to one of your questions, suggest ways to get around the difficulty, for example, "If you feel sick when you sit down, take a deep breath and think of something you really like to do."

Rewriting your child's script is no easy task. But it can set your kid up for success. Plan to incorporate imaging sessions into your coaching schedule.

The Trial Run

Having your kid set up an at-home pretest is an effective way to dispel test anxiety and reinforce positive thinking. But your child—not you—must design the test. Ask questions such as: "What do you think will be on the test?" "How will you know you 'passed'?"

Prompt your child to review notes taken in class, math homework, and entries in his or her learning log. Suggest that he or she define "success" as being better prepared or answering more questions. Try to dispel the idea that success means getting everything right.

Encourage your kid to set up the pretest on paper using questions and exercises from your at-home textbook. Then set a time to write the pretest. Your child will get the most benefit by working from memory. So be firm—no books or notes allowed.

> **Strategies for Test Day**
> - Get enough rest the night before.
> - Avoid cramming at breakfast for a morning test or at lunch for an afternoon one.
> - Prepare a test equipment package: sharp pencils, eraser, ruler, etc.
> - Scan the test first. Tick or highlight any questions you can answer. Do those first. Then go back and try more difficult ones. But move on to another question if you get stuck.
> - Focus on positive thoughts. Think "yes" instead of "no."
> - Plan a post-test reward.

Chapter 18

Celebrate Success—How to Turn Tests into a Positive Experience

This chapter refers to tests created by the math teacher or math department within a school. Students may also be required to write external exams, set and marked by a state, province, or school district. The results may be used to evaluate education in the tested region, and are often misused by the news media.

Self-appointed watchdogs bark when results are trumpeted in the news. Usually such tests are comparing apples to oranges. The results have very little to do with your child. It's important to remember that every test can become a positive experience, regardless of the mark your child receives. The key is to celebrate all successes, regardless of how small they might seem.

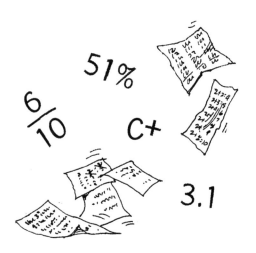

Why Do Teachers Give Tests?

From the teacher's point of view a test has two possible functions:

1. Some teachers give tests mainly to generate a number that can be placed beside a student's name on the class list. This number is combined with other numbers to create a mark or grade that will go on the student's report card. In some classes that is all any test is ever used for.

2. Better teachers use test results as part of their teaching program. Test results help these teachers evaluate their own teaching as well as their students' progress in mathematics. Good teachers want to know whether they have achieved the goals they aimed for.

Why Taking Up the Test Is Important

To ensure that students get maximum benefit from the test-writing experience, taking up the test is essential. In the military, this process is called "debriefing." In sports, it's "post-game analysis."

Good teachers build time into their course outlines and lesson plans for taking up the test. This process involves going through the test questions, providing answers, explaining the harder parts, and showing how answers were marked. This way the test serves a third function. It becomes a teaching tool.

In an ideal world, this post-game analysis should ensure that every child gets the picture. But kids who are having difficulty with math may be more confused than enlightened by this process—if it takes place at all. Increasingly a jammed math curriculum, crowd control, and other pressures eat into a teacher's time. So there is less and less opportunity for taking up the test and even less time to celebrate success. That's where you come in.

What to Do When the Test Comes Home

Knowing an answer's wrong is not enough. Kids need to recognize what mistakes were made and what misinterpretations caused the mistakes to occur. As math coach, you can fill in any gaps that remain when your child brings a test home. Start by praising everything that is remotely praiseworthy.

- Did your kid remember to bring the test home? That's terrific! Say so.
- Did your kid attempt every question? Most questions? More than on the last test? That's wonderful! Say so.
- Did your kid record and bring home any notes from the after-test discussion at school? That's phenomenal! Say so.

Your praise rewards your child's cooperation. It's sad but true that a kid who is failing math may not want to be the bearer of bad news. Trying the questions, taking down notes, and bringing the test home are signs that your child trusts you to be on his or her side.

At-Home Test Analysis

When the test comes home be ready to look through it and any notes your child has taken. Conduct a "sports interview" by asking: "What were you thinking when you saw the test?" "Read that question?" "Wrote this answer down?" (Review Chapter 12 for hints that will get your kid talking.) The next step is to help your kid frame questions to ask the teacher. The final step is to interpret the test results in order to plan future coaching.

The above process makes the test do extra duty. It lets you praise. It helps your child learn from mistakes. And it turns the dreaded math test into a positive experience.

What If the Teacher Won't Send the Test Home?

Some teachers won't send the test home. That's because they want to use it again next semester and they don't want copies circulating. This reluctance is understandable. But it can be overcome. Try these strategies.

- Ask the teacher to send the test home in a sealed envelope. Promise that it will be sent back the same way the next day, that no photocopy will be made, and that neither you nor your child will show it around.
- Ask if you and your child can review the test after school with the teacher present. (If you are normally at work after school, this won't be practical.)
- Explain (calmly!) why analyzing the test is important to you and your child. Ask the teacher to suggest how it might be accomplished.
- Turn the problem over to the guidance counselor. Don't give up!

Chapter 19

What if You Have to Turn Your Coaching Job Over to Somebody Else?

There are many good reasons why you might decide to turn your coaching job over to someone else. This chapter reviews your options and suggests how to get the best value without taking out a mortgage.

Video Tutorials

Without doubt video tutorials may help some kids some of the time. But we caution you: many misleading TV commercials suggest that you can somehow plug your child's head into the VCR and the math will flow in. That's a false hope unless someone makes sure your kid interacts with the tape by pausing, asking questions, and recording notes. Will that someone be you?

Videos do have one major advantage: they can be played again and again. But inevitably they fall into the "show and tell" category. If that teaching style really suited your child, she or he probably wouldn't be struggling in math.

Programmed Computer Tutorials

Programmed learning first came to prominence in the late 60s. The results were disappointing. Computerization has resulted in some improvement, especially for drill. But unlike you, computer programs can't recognize when your kid escapes mentally. Some newer school-based programs, in the hands of an experienced teacher, show promise in pinpointing a child's problem. But it still takes time and a warm body to correct the problem.

Learning Centers and Math Centers

The existence of learning and math centers shows that the demand for effective, affordable private tutors exceeds the supply. Just like private tutors, these centers have personalities. Some have a patented method. In their view, there's only one way to learn math. If that way works for your child, great; if not—tough. Other centers are more flexible.

If you've read the rest of this book, you should be able to evaluate whether a learning center gives good value for money. A good learning center is usually a less expensive option than a private tutor, but your child won't get one-on-one attention.

Private Tutors

Hiring a private tutor is usually the most effective option, but also the most expensive. Who's likely to be available? Moonlighting teachers. Retired teachers. College students. Senior high school students.

Ideally your child's school could recommend a tutor. But sometimes school authorities are reluctant to do so. Ask around—perhaps a friend or neighbor knows a good tutor. Place a notice in your church bulletin. Check the ads in your community newspaper. Place an ad yourself. But get references and make sure you follow them up.

Once you've chosen a candidate, agree on a trial period. Be very clear about how many hours you'll pay for (and at what rate!). Be aware that it can take any tutor several hours to zero in on a kid's problem. The work you've already done may save a great deal of the tutor's time, especially if your child has a learning log (see Chapter 15).

> Bob is a learning assistance teacher. He works in the learning center of a mid-sized middle school helping kids with math. Bob regularly encourages parents to help their kids and shows them how. Often a parent will say: "I never understood that concept." When that happens, Bob spends a few minutes explaining the concept to the parent. Nine times out of ten, the parent winds up saying: "You mean that's all there is to it?"

Peer Tutors

A peer tutor is a student volunteer who's the same age as your child or at most a year or so older. From a parent's point of view, peer tutoring has two main advantages: it's one-on-one and it's free. From a kid's point of view, peer tutoring's major advantage is that it cuts out adults. The major disadvantage is that a peer tutor may be a math whiz, but not a teacher or a coach. So he or she may not have the skill to shore up your kid's fragile "math ego." And peer tutors may not have your child's long-term welfare at heart. That's still your job.

Getting the Most from a Tutor's Time

Get the tutor to tutor you! Then when you coach your child, you'll be in a better position to understand what's going on. You already know your kid's learning style, so you will be able to be interpret for her or him. Even if you do hire a tutor for your child, being a well-informed math coach can reduce the tutoring time your child will need, and thus reduce the number of hours you will have to pay for.

Section 3
Coaching Secrets

Chapter 20   Breaking the Secret Code

Jemma's math book must be written in secret code! She's only eight, but she's supposed to understand "commutative!" I checked the course outline and sure enough, it says "commutative," too. Where can I get a decoder ring?   —Mercedes

We'll decode "commutative" in Chapter 21. But plain words may also have "secret meanings" that can keep your kid from succeeding. In this chapter we'll decode two of these plain words: "number" and "operation."

The Secret Meaning of Number

To mathematicians and math teachers, "4" is not a number. Neither is "IV" or "four." But "four," "4," and "IV" are **numerals**—names or symbols that represent a particular number. A numeral only represents —or stands in for—a number. The number itself doesn't "show." A number is the *idea* behind a numeral. This difference might seem too subtle to matter. However, many math curricula are organized around the idea that there are different kinds of numbers. But even the best textbooks seldom show clearly how the different kinds are related.

Take a look at the chart below. Think of it as a decoder ring to help

A Parent's Guide to Decoding the Language of Numbers

Kinds	Examples	Include	Don't Include	Comments
Natural Numbers	1, 2, 3 ...	"counting" numbers	zero numbers below zero	Natural numbers have been used for counting from earliest times.
Whole Numbers	0, 1, 2	zero and all natural numbers	numbers below zero	Whole numbers let us use zero to express huge values using only ten symbols (0–9).
Integers	...-1, 0, 1...	numbers less than zero all whole numbers all natural numbers	fractions percents decimals	Integers let us express numbers such as "below zero" temperatures.
Rational Numbers	-7 $-\frac{7}{1}$ 7 $\frac{7}{1}$ -½ ; ¼ 25% ¼ 0.25 ¼ 0.333̇ Ä	all fractions: ▪ common fractions ▪ percents ▪ terminating decimals ▪ repeating decimals all integers all whole numbers all natural numbers	nonterminating decimals nonrepeating decimals	Rational numbers can always be written or re-written as fractions with integers top and bottom. (Note how some examples in the second column have been rewritten as fractions.)
Irrational Numbers	p = 0.31416... square roots such as $\sqrt{3}$ that = 1.73205...	nonterminating, nonrepeating decimals	integers fractions square roots such as $\sqrt{4}$ that = +2 or -2	Irrational numbers *can't* be rewritten as fractions with integers top and bottom.

you interpret your child's math course outline. Read through the chart as many times as you need to. (Don't be discouraged if you don't "get it" in one read—decoding takes time. You don't have to memorize everything or cram it into your kid. Just use it to look up the terms you come across.)

The Secret Meaning of Operation

In math, an "operation" is a process that may affect the *value* or *form* of a number. Think of "value" as how much the number is "worth." For example, start with 2 and multiply by 5.
$$2 \times 5 = 10$$
This operation—multiplication—alters the value of the starting number (10 has a greater value than 2). Think of "form" as what the number looks like. For example:
$$4^2 = 4 \times 4 = 16$$
This operation alters the form of the starting number because $4^2 = 16$. So 16 is 4^2 in a new form.

There are four basic number operations. They are usually listed as addition, subtraction, multiplication, and division. The chart below uses a different order. Explore it before reading on.

> ### Confusion in the Operating Theater
>
> For kids, there are three major sources of confusion in performing number operations.
> 1. Operations can be described with symbols (+) or with key words ("plus"). But your child may get confused when more than one symbol or key word is used for the same operation.
> 2. Confusion also crops up when drill sheets and textbook exercises leave out the operation symbols. Kids facing something like this
>
> $$\begin{array}{r} 378 \\ \underline{45} \end{array}$$
>
>   often wonder if they should add, subtract, or multiply.
> 3. The third source of confusion arises from the traditional order for teaching operations:
>   *add, subtract, multiply, divide*
>   But this is not the best sequence for learning them. We say the best order is:
>   *add, multiply, subtract, divide*
>   In chapters 21 and 22 we'll show you why.

The Four Basic Number Operations

Operation	Symbol	Key Words	Answer	Examples
Addition	+	add plus	sum total	$13 + 6 = 19$
Multiplication	× · ( )	multiply find the product times	product result	$4 \times 5 = 20$ $4 \cdot 5 = 20$ $4(5) = 20$
Subtraction	−	subtract find the difference minus	difference result what's left	$19 - 3 = 16$ $13 - 6 = 7$
Division	÷ / $\overline{)}$	divide find the quotient divide by divided by divide into	quotient result	$21 \div 7 = 3$ $21/7 = 3$ $\frac{21}{7} = 3$ $7\overline{)21}^{\,3}$

Chapter 21

The Secret Behind Addition and Multiplication

Even the earliest humans used addition. It's a simple-seeming idea—put like things together to find out how many there are altogether. The idea of "like" is important. You can add 1 cow and 5 calves to get 6 cattle. But you can't add 1 elephant and 5 eggs to get 6 elephant eggs.

Multiplication Is Accelerated Addition

Many years ago, Louise was buying a set of 12 mugs at 59 cents each. You can see on the left how the clerk worked out the total. She wasn't confident of her multiplication skills, so she added.

Multiplication is the natural next step after learning addition. But most kids are taught subtraction right after addition. By the time they meet multiplication, many kids already have a head full of subtraction rules and difficulties.

When children first learn to multiply they often use pictures or manipulatives to show that multiplying means adding like groups. Soon, however, kids learn to use times tables to multiply two numbers with single digits. Then they learn how to use the method shown below to multiply two numbers with two or more digits, which is much faster than adding the same number over and over.

Here's what your child is really doing: *multiplying* digits one pair at a time, placing the answers in columns, and *adding*. But for a kid who's struggling, doing all this in the right sequence can be exhausting. It gets hard to picture *multiplying* 12 × 59 as *adding*

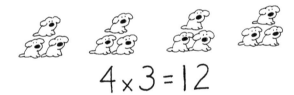

12 groups of 59. The connection gets lost—it becomes a big secret. If your child is struggling with multiplication, you may have to go back to the first principles of addition. Don't be afraid to use pictures, beans, paper money, etc. And *do* impress on your child the value of learning the times table. (Chapter 24 suggests how you can blot out the times table blues.) Having those facts on tap in mental storage can go a long way toward building self-esteem.

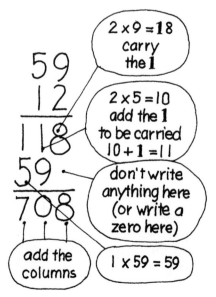

Properties of Addition and Multiplication

"Properties" are rules that affect how addition and multiplication can be used to manipulate numbers. Here are some important ones.

Addition and multiplication are *both* associative

Associative refers to grouping and regrouping numbers. Look at this picture. Does regrouping change the number of cookies?

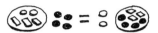

Here's the same picture in symbols: $(2 + 3) + 4 = 2 + (3 + 4)$

For addition, parentheses act like plates to group numbers together. For multiplication, however, parentheses may be used instead of the × symbol. Look at this picture:

Here it is in symbols: $2(3) + 2(4) = 2(3 + 4)$.   Again regrouping left the number of cookies unchanged. So "associative" means that regrouping numbers is okay when adding or multiplying.

Addition and multiplication are *both* commutative

"Commutative" refers to order or sequence. In this addition example, the sum is still 11, whether the 5 comes first or the 6.

  $5 + 6 = 11$ and $6 + 5 = 11$ so $5 + 6 = 6 + 5$

Commutative applies to multiplication, too. In this example, the product is still 30, whether the 5 comes first or the 6.

  $5(6) = 30$ and $6(5) = 30$ so $5(6) = 6(5)$

So "commutative" means that resequencing numbers is okay as long as you're adding or multiplying.

Associative Property	
Addition Regrouping numbers to be added does not change their sum.	**Multiplication** Regrouping numbers to be multiplied does not change their product.
Commutative Property	
Addition Changing the order of numbers to be added does not change their sum.	**Multiplication** Changing the order of numbers to be multiplied does not change their product.

Only multiplication is distributive

It's Halloween. There's a costumed group of two boys and three girls on the porch. You distribute four wrapped gumballs to each kid. Here's how that looks in symbols:

  $4(2 + 3) = 4(2) + 4(3)$

The 4 is a multiplier. It applies to each and every term inside the ( ) because multiplication is "distributive." This means applying something "equally" to each "item" in a group like this:

  $4(2 + 3) = 4(2) + 4(3) = 8 + 12 = 20$ or $4(2 + 3) = 4 \times 5 = 20$

The distributive property applies *only* to multiplication. It can't be applied to addition.

In this chapter you have seen that addition and multiplication are strongly linked to each other. In Chapter 22 we will show you that subtraction and division also have a close relationship.

Chapter 22   The Secret Behind Subtraction and Division

*James was brighter than average and seemed to be doing well in math until he met algebra in grade 9. My testing revealed a weak grasp of subtraction. I wasn't surprised. That concept often lies at the bottom of math problems. But it doesn't really **hurt** until kids start algebra. By then, they're adolescents. I have to be very courteous when I explain that "the cure" involves doing "grade 3 math."*   —Tina, math tutor

Stumped by Subtraction

Although almost all early cultures knew some form of addition, many did not have subtraction. Why? Because it's a tough concept. Visualizing addition is much easier than visualizing subtraction. Look at this pencil picture.

What's the difference between the groups? Most adults would think 4 – 2 = 2. But Jean has tutored kids who would say: "The *difference* is 1, because if you take away 1 pencil from the big group and give it to the small group then both groups will have 3 pencils. They'll be the *same*."

Another language trap! Kids learn early that "different" is the opposite of "same." There are so many other ways of asking a subtraction question: "What's the difference?" "How many more or less?" "Bigger or smaller?" "By how much?" No wonder it's hard for kids to recognize when or what to subtract!

Worse still, subtraction is often taught before addition is firmly grasped. A child who's not ready for subtraction can often recite facts so well nobody realizes there's a conceptual problem.

Algorithm, Anyone?

An algorithm is a set of steps used for calculating. The word is common now because computers use algorithms. But there is nothing new about the idea.

Algorithms save time, but at a price. Most subtraction algorithms actually block understanding. According to recent research, our minds do subtraction along a mental number line. Unconsciously, we visualize differences as distances. Explore the Milestone Algorithm at left. It gives the right answer, is easy to use, and retains the true nature of subtraction. Maybe it can help your child.

Milestone Algorithm

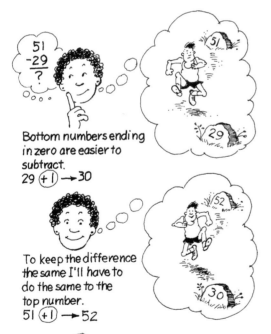

Bottom numbers ending in zero are easier to subtract.
29 (+1) → 30

To keep the difference the same I'll have to do the same to the top number.
51 (+1) → 52

51 (+1) → 52
29 (+1) → 30

The answer is 22

Done in by Division

If subtraction is tough, division is tougher. Here's why:

1. You can't easily memorize "division tables" although strong times table skills are helpful for "short" division. So 20 on the times table:

    $20 = 4 \times 5$, so $4\overline{)20}^{\,5}$

2. Division is often said to be the "opposite" of multiplication. But this focus conceals an important relationship: division is just repeated subtraction. So you can't understand long division until you understand subtraction (although algorithms may help you do both without understanding either). Look at the algorithm on the right and you'll see that it's true.
3. This algorithm involves a repeated cycle of estimates, trial multiplications, and trial subtractions—all in the right order. Brain research confirms that it's just as complex as it looks.
4. There are several different division symbols in common use:

    $20 \div 4$ or $20/4$ or $\frac{20}{4}$ or $4\overline{)20}$

5. There are so many ways to state a division problem: *divide; divide into; divide by; go into,* etc. No matter how well you know the algorithm, you can't get the right answer unless the terms are in the right order. That's just as true if you use a calculator: dividing 4 pizzas among 20 people gives a much different outcome from dividing 20 pizzas among 4 people.

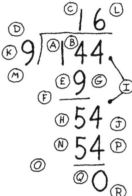

As if that Weren't Enough…

Sometimes it seems that every English-speaking jurisdiction in North America expects kids to study long division in grade 4. Yet reliable research shows that only a quarter of kids this age can actually understand long division or even use an algorithm to perform the operation consistently and correctly.

Don't be dismayed if your child has trouble meeting those expectations—they may be unrealistic. You may need to rebuild shattered confidence by rebuilding basic math skills. Let your child know that he or she is not "stupid"—it will be okay eventually.

A 9 into 1 "won't go."
B 9 into 14 goes 1 time.
C Write 1 above the 4 in 14.
D Multiply that 1 by the 9 (1 × 9 = 9).
E Write the answer (9) under the 4 in 14.
F Draw a line under the 9.
G Subtract the 9 from 14 (14 − 9 = 5).
H Write the answer (5) under the 9.
I "Bring down" the final 4 from 144.
J Check: The 54 that's left still hasn't been divided.
K Divide 54 by 9 (54 ÷ 9 = 6).
L Write the answer (6) above the final 4 in 144.
M Multiply that 6 by 9 (6 × 9 = 54).
N Write the answer (54) under the original 54.
O Draw a line underneath.
P Subtract (54 − 54 = 0).
Q Write the answer (0) under the 4 in 54.
R Check: The 0 means there's nothing left to divide.

Chapter 23

The Secret Behind Calculators

At the supermarket today, the register said I owed $23.37. "That's way too much," I told the cashier. "It should be around six dollars." He didn't get my meaning at first. But I stuck to my guns—I had three items—$1.93, $1.79, and $2.25. So I knew I was right. When he printed out the bill, it was easy to see the problem. He'd punched in $19.33 for the soup bones, instead of $1.93. When he re-punched the sale, the new total was $5.97. He was amazed. "Nearly six dollars! How did you do that in your head? What's your secret?" —Beatrice

Mental Arithmetic—A Survival Skill

Many people assume that calculators have eliminated the need for arithmetic skills. This is definitely not true. Here are three strong reasons why every kid needs arithmetic skills.

1. Survival in a computerized consumer society

Despite computers and calculators, you can still wind up being overcharged. It isn't fraud. It's sheer complexity. To protect yourself and your pocketbook, you need to be able to estimate.

2. Increased self-esteem and confidence

Reliance on calculators makes us distrust our own judgment. Someone less confident than Beatrice might have paid the $23.37 and left the store feeling victimized.

3. A foundation for future math studies

Having basic arithmetic facts in mental storage with no need to use a calculator will give your child an edge in advanced math such as algebra. Your kid's mind will be free to cope with algebra concepts, while others are still punching a calculator or struggling to recall those "boring old baby facts."

Estimation—An Essential Survival Skill

Remember the sales clerk on page 54? Beatrice would have had the right change out before the clerk wrote the second 59. Here's how:
Rounding off ➜ 59 cents is just 1 cent less than 60.
Mental arithmetic ➜ 6 × 12 = 72, so 60 × 12 = 720.
Estimation ➜ 12 mugs at 60¢ each should cost around $7.20.
Judgment ➜ So twelve 59¢ mugs should cost a little less than $7.20.
You don't need to find out the exact answer. Estimation means getting close enough to judge whether the calculator is in the right

ballpark. To do this, your child must have some key arithmetic facts in mental storage.

Did the Robot Flunk the Math Test?

Most kids who rely on a calculator for routine test calculations finish last and do poorly. Why? Because the very calculator features we trust most—speed and accuracy—can't be completely trusted.

Kids who use a calculator for every test calculation very often don't finish the test at all. That's because kids are human. In the time it takes a human to punch a simple problem like 5 × 16 into a calculator and read the answer, a child with effective mental arithmetic skills can complete two more questions.

Calculators have no secret powers. They aren't smart. They can't judge. A calculator can't tell if a finger twitched and entered an extra 6, or if a finger slipped and missed a 5, or a decimal sneaked into the wrong position. The robot in the calculator will do exactly what it's told—no more and no less.

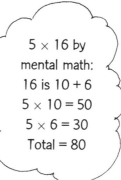

5 × 16 by mental math:
16 is 10 + 6
5 × 10 = 50
5 × 6 = 30
Total = 80

Cast Off that Helpless Feeling

Tasha, Jasmin, and Celeste are writing a science test. In Question 1 they need to multiply 5 by 16. Tasha and Jasmin use their calculators. Celeste knows her brain is just as fast or faster—so she uses mental arithmetic. She's probably never seen 5 × 16 on a times table, but her secret is to break 16 into an easy-to-multiply equivalent—10 + 6—at a single glance.

In Question 2 they need to multiply 0.86603 × 5280. All three use a calculator. Tasha gets 45726384, Jasmin gets 457.26384, and Celeste gets 4572.6384. Celeste uses estimation to check up on her calculator. She has the time she saved on Question 1 to do this.

Rounding off → 0.86603 is about 0.9 (rounding up)
Rounding off → 5280 is about 5000 (rounding down)
Mental math → 0.9 × 5000 = 9 × 500 = 9 × 5 × 100 = 4500
Judgment → 4500 is close in size to 4572.6384

Celeste moves on, confident that her calculator's answer is in the right ballpark. Tasha and Jasmin move on too, but they *don't* know their answers are way out in left field. The mental math secret here is recognizing that 9 × 500 has the same value as 0.9 × 5000 and is easier to multiply. It's as if 0.9 × 5000 were a "disguise" for 9 × 500. In Chapter 24 we'll show you how to recognize numbers in disguise.

Chapter 24 Secrets for Re-energizing Drill

Nothing provides more of a boost to a kid who's struggling with math than being successful. Here's how Ginnie helps her middle school students get that feeling when they come to the math center. She gives the kids the same drill sheet day after day. After a few days of getting them all right, students will say: "Gee, Ms. Eyres, you know I haven't really learned these. I just memorized them."

"Gee," says Ginnie, "you sure fooled me!"

The kids think they are getting away with something, when in fact Ginnie has "psyched" them into memorizing all sorts of math facts. Once the facts are in mental storage, the kids are confident to use them in new situations.

Beating the Times Table Blues

A kid who's struggling needs a custom-made times table. You can help make it, but your child must decide on the design.

Use highlighters to make the columns and rows easy to follow. Make the numbers big enough to read from two to three feet.

Standard times tables usually go up to 10×. But a customized table should include 11× (it's easy), 12× (eggs and pens come in dozens), 100×, 1000×, and 1 000 000×. Your child (not you!) must fill in the boxes, but it's all right to use a calculator for this job.

Whenever homework involves multiplying, prop the times table up and let your kid use it freely. To your child, it may be a welcome crutch. In reality, it's a subtle form of drill that reduces anxiety. Soon you can ask your child to guess the answer aloud before looking at the table. Guessing right provides a tremendous boost. Your kid can also use the table to explore the anatomy of numbers.

×	0	1	2	3	4	...
0	0	0	0	0	0	
1	0	1	2	3	4	
2	0	2	4	6	8	
3	0	3	6	9	12	
4	0	4	8	12	16	
⋮						
12	0	12	24	36	48	

Anatomy of a Number

Pick a number—any number. *What's it made of?* Few kids realize that it is made of anything. But every whole number except 1 can be dissected to reveal its hidden parts. Consider the anatomy of 6:

$6 = 5 + 1$ and $6 = 2 + 4$ and $6 = 3 + 3$ and $6 = 2 \times 3$

Dissecting a number to uncover its secret disguise is an extremely useful exercise for your drill sessions. Here are two dissection tools to help your child build number anatomy charts.

1. Backward multiplication—a key dissection tool

For younger kids, you can call this dissection tool "backward multiplication." For older kids, its official name is "factoring."

Let's use 18 to illustrate the method. First scan the times table to find every product box that contains 18—you'll find it in at least two places depending on the design of the table. Then track backward through the table to find different ways to "build" 18 using multiplication.

$$18 = 2 \times 9 \text{ and } 18 = 3 \times 6 \text{ and}$$
$$18 = 6 \times 3 \text{ and } 18 = 9 \times 2$$

Here are two other ways to build 18. They may seem "too obvious" to mention, but teachers insist that kids know them.

$$18 = 1 \times 18 \text{ and } 18 = 18 \times 1$$

Now begin assembling an anatomy chart for 18 using all those disguised parts you have uncovered.

```
      3 x 6     6 x 3
   2 x 9  \   /  9 x 2
          — 18 —
   1 x 18 /   \  18 x 1
```

2. Backward addition—another dissection tool

Backward multiplication (factoring) is only one way to dissect 18. Backward addition reveals more hidden parts such as:

$$18 = 1 + 17 \text{ and } 18 = 2 + 16 \text{ and } 18 = 3 + 15$$

and so on all the way to:

$$18 = \boxed{15} + \boxed{3} \text{ and } 18 = \boxed{16} + \boxed{2} \text{ and } 18 = \boxed{17} + \boxed{1}$$

Notice that as the circled numbers go up by 1, the boxed numbers go down by 1. That's a pattern. Expand the anatomy chart for 18 by adding the parts of this pattern.

```
         3 x 6     6 x 3
      2 x 9  \   /  9 x 2
             — 18 —
      1 x 18 /   \  18 x 1
```
... 15 + 3 ← 16 + 2 ← 17 + 1   1 + 17 → 2 + 16 → 3 + 15 ...

Familiarity with anatomy charts can give your kid a real taste of success. If a child can see 18 and immediately identify 13 + 5 or 9 × 2 then the answer to 18 – 13 or 18 ÷ 9 is easy and automatic.

Getting It All into Mental Storage

For most strugglers, physical storage promotes mental storage. So help your kid choose or make a container to use as a personal math "toolbox." Add some large envelopes and you're ready to begin.

1. Collection: Build a "tool collection" that's meaningful to your kid. Be sure to establish appropriate rewards for starting each new chart, table, or card. Your child's own sense of organization must prevail. Don't insist on adding facts that aren't needed yet.
2. Repetition: The act of making a tool and storing it in a labeled envelope may be enough to store it in memory. If not, don't worry. When the fact is needed, let your child get it out and use it. In time, repetition will transfer the fact into mental storage.
3. Selection: You can't tell in advance which tools are going to be most useful. Keep the ones that are used frequently at the front of the box. Dating the envelope each time a tool is used may help your kid decide which ones can go to the back.

Selected Resources

Human Resources

Family Counselor
A child who can't recognize the value of planning for his or her own math future and acting on that plan might benefit from counseling outside school. Get a referral from someone you trust.
Physical Therapist
A child who gets totally stressed out before and during every math test could benefit from learning relax-on-demand techniques such as those that professional athletes and actors use.
Physician or Pediatrician
Your child's regular MD can decide if any health issue may be affecting math performance.

Reading Resources

The 7 Habits of Highly Effective People by Stephen R. Covey ISBN 0-671-70863-5
Not just for sales managers—this book discusses effective ways to collaborate with your kid even while asking him or her to work harder. Inexpensive. Go to the index for emergency help.
The Math Curse by Jon Scieszka ISBN 0-670-86194-4
Written for kids. Most libraries have this witty book.
Overcoming Math Anxiety by Sheila Tobias ISBN 0-393-31307-7
Written for adults. If you're harboring math anxieties of your own, this one's for you.
Science for all Americans by F. James Rutherford and Andrew Ahlgren ISBN 0-19-506771-1
Written for the general reader and just as useful to Canadians. Don't let the title fool you. Chapters 2, 9, 12, and 13 have much to say about math. Inexpensive.
Benchmarks for Science Literacy Project 2061 ISBN 0-19-508986-3
Published by the American Association for the Advancement of Science. Written for educators, but remarkably readable. See especially chapters 2, 9, and 12. Expensive—ask at the library.

Curriculum Resources

Official curriculum guides are written for teachers, but some parts may be practical for parents.
Math On Line: Cyberspace is vast and uncharted. Key words such as mathematics, curriculum, guide, framework, may yield 100 000 plus "hits," but many are duplicates, useless, outdated, obscure, or eccentric. Your time may be better spent coaching your kid than sorting web sites.
Surf Alaska: Alaska's web site is an exception—it's the most accessible we've seen. Ignore the "math-ed" jargon up front. Scroll through to the Big Ideas. They'll help you assemble a coherent picture of school math. <http://transition.alaska.edu.www/DOE/Mathsci/ms3cntn1.htm>
Prefer Print? There are hundreds of distinctly different math curriculum guides or frameworks in use in the US—far too many to list here. But Yes You Can! Press maintains an index of the main ones, and you can get it at little or no cost. Read about it on page 63. To order, see page 64.
Phone Wisconsin: This state's curriculum documents are exceptional. The math guide goes far beyond what-to-teach-when-and-how, so parents from any state or province may find it worth braving the mathematical and educational jargon. To order, call toll-free 1-800-243-8772.
Canadian, Eh? Order the current curriculum guide or framework for your province from Micromedia Ltd., 20 Victoria St., Toronto, ON M5C 2N8.
Prefer Something Simple? Our Generic Math Curriculum Grid covers Kindergarten to Grade 9—in the plainest language possible. Read about it on page 63. To order, see page 64.

Yes You Can! Help Your Kid Succeed in Math Even if You Think You Can't

How to Order Additional Copies

By phone...

Toll free 1-888-YES-PRESS That's 1-888-937-7737.
It's toll free everywhere in the United States and Canada.

On the web...

Go to: yesyoucanpress.com
and follow the easy instructions on the home page

By mail...

In the USA	In Canada
Yes You Can! Press	Yes You Can! Press
2400 NW 80th Street PMB 173	1511 Marine Circle
Seattle WA 98117	Parksville BC V9P 1Y5

Enclose a check or money order made out to *Yes You Can! Press*

$18.00 (for the book) + $4.00 (for shipping)   plus applicable taxes

If you are not sure how much tax to include, call 1-888-YES-PRESS

Or...

From your favorite independent bookstore. They'll need this information:
- title: Yes You Can! Help Your Kid Succeed in Math
- author: Jean Bullard and Louise Oborne
- publisher: Yes You Can! Press
- ISBN number: 0-9658044-0-2
- phone number: 1-888-YES-PRESS (1-888-937-7737)

Selected as "Outstanding" by the Parent Council® Ltd

"A practical and helpful resource, especially for those parents involved in remedial, home enrichment, or home schooling activities. This book explains why a parent is the best math coach a child can have, regardless of parental education level or math ability. The authors use plain language to penetrate "math mystique" and offer twenty-four accessible, helpful chapters organized into three sections: Pep talk, Winning Strategies, and Coaching Secrets." (www.parentcouncil.com)

Recommended by the Davis Dyslexia Association

"Advice for parents and strategies for overcoming math anxiety and other barriers to learning. A great resource for those who don't know where to start." (www.dyslexia.com)

What Others are Saying About *Yes You Can!*...

- **Travis Cavens, M.D., Board-certified pediatrician**
  You two have created a marvelous book. It is a book that any pediatrician can recommend with confidence to parents whose children are struggling with math.

- **Keith Clark, math consultant**
  What an excellent book! It really goes to the heart of helping students with mathematics—namely the role of the parent as coach. The suggestions are clear, logical, and practical.

- **Ian de Groot, award-winning teacher and NCTM[1] committee member**
  I think it's a very effective book. It tells you, if you really want to help your kid, what you have to do… It does cover every possible base: textbooks, study habits.

- **Barbara J. Dolan, speech language pathologist and parent**
  I am one of the math "dumbbells" and to this day I actually "go blank," then panic sets in and I "go blanker." However, I started reading as I promised, and to my great surprise I found myself becoming more and more interested and actually understanding some of the math concepts. From a parent's point of view: I feel as if I could and would be able to help my child. From a professional's point of view: this book supports all the principles of positive parenting.

- **Ron Lancaster, award-winning teacher and NCTM committee member**
  Parents who are frustrated with their child's progress in mathematics would do well to take a close look at this book. It will empower parents to do something about their child's education in an area where they do not often feel comfortable. The many suggestions have been well thought out.

- **Jane Hurley Davis, award-winning math teacher**
  You two have addressed the issues that face many parents and students. Confidence leads to success and this book gives appropriate guidance.

---

[1] NCTM: The National Council of Teachers of Mathematics is an American-based organization with thousands of members throughout the United States and Canada. Its views have strongly influenced mathematics education in all 50 states and 10 provinces. The views expressed above are the opinions of the individuals quoted. No claim is made with respect to the NCTM.